Circumnavigations
The Shrinking of Our World

Alan Boone

Copyright © 2012 Alan Boone

All rights reserved.

ISBN:1491227958
ISBN-13:9781491227954

DEDICATION

To Sandy, who believed in this work

Contents

Preface 7

Chapter 1 - Del Cano (Magellan) 9

Chapter 2 - Drake 31

Chapter 3 - Cavendish 47

Chapter 4 - Shouten-Lemaire 55

Chapter 5 - Roggewein 62

Chapter 6 - Byron 67

Chapter 7 - Marchand 74

Chapter 8 - The Clippers 80

Chapter 9 - Cochrane 90

Chapter 10 - Also Rans 102

Chapter 11 - Ekins 110

Chapter 12 – Gallagher 120

Chapter 13 - Olds 128

Chapter 14 - Gagarin 132

Circumnavigation Record Speeds 138

Index 139

Bibliography 146

PREFACE

In the peaceful silence of space, a vehicle turned in preparation for re-entry into the earth's atmosphere. Inside the pressurized cabin, a single passenger rode in cramped but relative comfort over 25,000 miles around the earth at an altitude that reached over 100 miles. He had been less than two hours in route.

Over four hundred years earlier, the first men to circle the globe struggled across its watery surface for nearly three years before returning home. Only 18 out of 277 men made it to harbor on a historic day in 1522. In a real sense, they set the first speed record for a circumnavigation of the world. How long did that record stand? How did mankind shrink its world from three years to two hours during the course of five centuries?

The story is a tale of transportation itself. It is the story of man, technology, and politics in a changing world. Uncovering these adventures is a quest which took me from the encyclopedia, to the library, and finally through the centuries; the age of exploration, the clipper ship and early steamship eras, dirigibles, propeller driven aircraft, jets, and rockets.

Thousands of circumnavigations have been made by many means, and it is so common that most are not recorded. For purposes of my search, I defined a circumnavigation as leaving and arriving at the same longitude or farther after circling the globe and travelling more than 22,858.74 miles, the distance of the Tropic of Capricorn. This is the distance criteria used by the Federation Aeronautique International, and it leaves out many fast circumnavigations that chose a much smaller circle around the upper part of the Northern Hemisphere. A true circumnavigation should also pass through two antipodal points which are at least 12,429 statute miles apart.

The project would probably have died in infancy, were it not for the early discovery of a thorough and scholarly book by Robert Silverberg entitled, *The Longest Voyage*. This yielded up the first four records. Other leads were found tucked away in obscure paragraphs of the books listed in the bibliography and the present list took shape. Newspapers and museums filled in many blanks, and finally my wife who encouraged the project, and my daughter who insisted that I provide sketches and maps to help with visualization.

None of this search would have been possible without the tireless efforts of the libraries and museums around the world that are dedicated to the preservation of the heritage and collective knowledge of mankind. We also owe much to those early travelers that kept records of their journeys. They have now become our journeys.

Although I believe this list to be complete, there may be other record breaking globe-trotting events that I have missed. If there are, I would be interested in putting each new puzzle piece in its place. My hope is that someone interested in the subject will find in this one volume the answers to questions so laboriously researched, or that others can take this work and further answer questions that pop into the curious and slightly romantic mind.

CHAPTER 1- DEL CANO (MAGELLAN)

Five tiny ships crowded with 277 men weighed anchor in a Spanish harbor in 1519. The first assault on the round world had begun, though few of the eager crew members would live to see its conclusion. By the time they returned, the world would have expanded to its largest size, for only then would men know of the vastness of the planet they called home, and especially the immense Pacific Ocean. Thousands of years of human history set the stage for this event.

Long centuries before Christ, Egyptians and Phoenicians, Greeks and Carthaginians had mastered the Mediterranean Sea and established profitable trade and travel. In some of their travels, they ventured as far as the northern coast of France, and, according to the Greek historian, Herodotus, a Phoenician fleet travelled around Africa at the behest of the Egyptian Pharaoh Necho, around 600 B.C. Greek mathematician Pythagoras believed that the earth was a sphere, and Eratosthenes made the remarkably accurate estimate that the circumference was 25,000 miles.

After the fall of Rome, much knowledge was lost, and there was little success in restoring that knowledge until the Renaissance. Science was frequently molded to fit current religious dogma, and precious knowledge of the earth disappeared from Western thought. The old concept of an earth with five zones was put forth in the eighth century. This divided the world into a Northern and Southern Zone. These would correspond to the Arctic and the Antarctic areas. They were considered uninhabitable due to extreme cold. There were two Temperate Zones where everyone lived, and in the middle was an uninhabitable Torrid

Zone, corresponding to the Equatorial areas. For many years, among Europeans there was a fear that no one could survive the crossing of the Torrid Zone.

Meanwhile, Arab sailors wandered throughout the Eastern world establishing a rich and flourishing trade with India and China, and rapidly developing the technology of seafaring.

Need and avarice had forced the Europeans out of their restricted habitation. Lacking sufficient fodder to carry their livestock through the winter, they had to slaughter much of their livestock in the fall of each year for the rest of their herds to survive. The food supply of Europe during the winters depended on spices to preserve the meat for the seven or eight lean months of the year. Pepper, cloves, cinnamon, and nutmeg were valuable and could produce handsome profits for the venturesome trader. Overland caravans carried rich cargoes of silks, spices, gems, incense, and jade from China. Arabs brought the goods by sea and caravan. Europe offered wool and linen textiles, metals from their mines, armor, weapons, gold and silver. The riches of the East brought a profit at each stage of the journey from China, India, and the Spice Islands. Profits accrued to the growers of these lands, to the shippers, to the Arabs, and finally to the Italian merchants who distributed the vital goods to the rest of Europe through Venice and Genoa. Cargoes were frequently decimated by pirates from China and Malay lurking in the Java Sea. Any new route could bring great profit to the country and the man undertaking the voyage.

Two countries, Spain and Portugal, pushed back the boundaries limiting the countries of Europe. In the early 1400s, Portugal, with the help of the talented scholar, Prince Henry the Navigator, slowly struggled its way around Africa. Finally, she had discovered the sea route around the southern tip of Africa named "Cape of Good Hope" by the King of Portugal, but "Cape of Storms" by the sailor who discovered it. This opened the way to India and the Moluccas for the Portuguese, along with a brutal war between them and the Moslems; who felt the intrusion into their territory. Portugal had just come through a centuries long battle to drive the Moslems from their territory, and the animosity ran deep.

Spain had opened the door to the West in 1492, when Christopher Columbus crossed the Atlantic, and ran head on into a new land that seemed to stretch endlessly from pole to pole blocking the way to the

Indies. Only the most superficial beginnings of the exploration of the New World had begun. Balboa crossed the Isthmus of Panama in 1513, and viewed the vast Western Ocean. Then the battle for sea routes to the Indies had taken shape, the Portuguese going east and the Spanish going west.

An attempt was made to settle the issue with the highest political authority, the Pope. In 1493, a Papal decision divided the newly discovered lands between the two rival powers. As this New World began to take shape, and with neither country knowing the extent of these continents nor of the ocean beyond, treaties and additional Papal decisions attempted to clarify the issue. By 1514, Pope Leo X had endorsed the view that the line of demarcation (which divided the territories of Spain and Portugal) ran 46° west and continued around the world, becoming 134° east in the pacific. This gave Portugal a foothold in Brazil, and Spain an unknown portion of an unknown and apparently unprofitable continent. Yet beyond that lay the riches of the East, and Spain hoped that some of these islands lay in their territory and could still be found by sailing west and finding a way through the pesky New World.

The world of 1519 was in the midst of an orgy of new discovery, a world torn by greed and politics, a world of vastly different cultures coming together in a clash of European commercial rivalry, religious zeal, and exploitation. Europe was under the powerful political force of the Catholic Church, the Middle East was held under the equally powerful religio-political force of the Moslems, China was aloof, and Polynesia and Japan were isolated. The African continent was mostly unknown and a recently discovered New World to the west seemed to stretch a vast and formidable barrier to Europeans trying to gain access to the riches of the East.

The art and science of travelling across the earth's seas had also developed over the centuries. The ancients built ships of ever increasing size, propelled by oars and aided by a square sail when the wind was right. By the mid-15th century, while oar driven galleys still roamed the Mediterranean, Prince Henry hastened the improving technology of seafaring. He improved cartography, and he worked to improve the quadrant (an instrument for measuring the height of the sun or a star to determine latitude), the astrolabe (a more sophisticated version of the quadrant), and the compass.

Ships now had stern rudders and triangular lateen sails that allowed them to tack, sailing at angles against headwinds. The well-developed Portuguese caravel was well suited to the explorers, and opened the way around the southern tip of Africa to the East for that country.

Little is known of the five ships which began the first fateful cruise around the world, although much study has been devoted to the ships of Columbus, which crossed the Atlantic 27 years earlier, and the Golden Hind, which was to follow these ships around the world a half century later. We know that Magellan's ships were old and small and probably not unlike the vessels of Columbus. These ships were generally broad beamed, capable of carrying a profitable cargo home, and remaining stable in rough seas. The rigging was simple compared to future standards and may have been two-masted with some combination of square and lateen (triangular) sails. For years the sailor would live in this wood, rope and rigging environment with little protection from tropical sun and Antarctic cold, and only the rolling unsettled sea for viewing. He could sleep on deck or below in the newly invented "hammock", and only the officers could have a cramped and sparse cabin. Speed was not the forte of these sturdy little ships, but at least one managed to wallow the distance around the globe. It remained for a man of resolute will, tested seamanship, strong leadership ability, and a dream to coax the ship far enough for another to take over and complete the trip.

Ferñao de Magellanes, known centuries later in English speaking countries as Ferdinand Magellan, stood on the deck of the Trinidad on September 20, 1519. It seemed that his whole life had been in preparation for this moment. He was nearly forty, a veteran seaman and a veteran soldier, having been wounded in the service of Portugal during its bloody conquests of the Moslem trading centers of the East. Tough, cold, calculating and skilled, he had managed, through a combination of luck and determination to get backing for his bold venture.

Never part of the in-crowd in Portuguese politics, he was considered a traitor by some. After being insulted by Dom Manuel, King of Portugal, he took his loyalty, and his dream elsewhere, to Spain's King Charles. A prophet is not without honor, save in his own country and Magellan found open ears in young King Charles.

He had carried with him the astronomical and geographical expertise of his friend and confidant Ruy Falliero, who had originally talked him into going to Spain. Together they agreed not to reveal what they believed to be their secret, the westward route to the Spice Islands, those coveted lands of wealth. Magellan already had a friend in those islands who wrote to him of the riches that were ripe for exploitation.

Falliero and Magellan were wrong, but very persuasive. What they probably had was an incorrect map showing the mouth of the Rio de la Plata as a hole in the new world, and a completely mistaken concept of the size of the earth. It was an idea sufficient to convince King Charles, however, and a private investor, who rounded out the capital requirements necessary to fit out such an ambitious undertaking. Together, they put up 8,334,335 marevedis, over ¾ of which was advanced by His Majesty. His was a handsome investment in a risky project, but it was an era when such gambles could pay off richly if it resulted in a shipload of spices brought to market or a sovereign claim on a new land.

Magellan had worked diligently to assemble ships and crew for over a year. Supplies, ordinance, and bartering goods had to be gathered. A crew had to be assembled. Magellan's efforts were tireless and meticulous. The five ships had to be purchased, transported to Seville, caulked, and repaired.

These were still happy times for the man who had been slighted by his native Portugal. His work was frustrating and difficult, but he now had his dream, his command, and a new wife who had presented him with a newborn son. She now carried a second child that he was destined never to see. It was a time of friends, family, and a burning goal, and he must have been sorry to see these days come to a close, even though he longed to fulfill his destiny.

His crew was a motley group of Spanish, Negroes, Dutch, one Englishman, Germans, French, Italians and as many Portuguese as the Spanish backers would allow. A tough group of sailors, they were willing to sail off to anywhere for advance wages. The ships' captains were Spanish noblemen.

(Authors note: Negro, from the Latin word for black, was a Portuguese term for the dark complexioned peoples of Equatorial Africa. It was not until the 1970s that the term came to be considered to have any negative connotation.)

Among the group with Magellan were three other men of destiny. Magellan's slave, Enrique, would actually be the first man to return to his native area after travelling around the globe. Brought to Europe by Magellan from his earlier trips to the East many years before, Enrique would find a home land in the Philippines before the Europeans could journey on to theirs. If there were a record of his return to his starting point, he would be the first man to circumnavigate the globe. Another man was young Juan Sebastian Del Cano, destined to captain the survivors back to Spain. The third was an Italian adventurer, who went not as a crewmember, but as a tag along and chronicler. It is from the journal of this man, Antonio Pigafetta, that much of the knowledge of this remarkable voyage is derived.

The fleet of five ships that sailed out to sea was provisioned for a voyage of two years. They were laden with what was needed for ships, men, and trade. For the ships there were spare sails, beams, planks, pitch and tar, spikes, nails, block and tackles, canvas, lanterns, tools, flags, rope and small boats. Weaponry was vital in an age of international rivalry, and exploring uncharted lands. The fleet carried suits of armor, helmets, breastplates, spears, arquebuses, powder and shot for the men, fieldpieces, falconets, and three large bombards as cannons for the ships, along with molds for fashioning cannonballs. Staples included biscuit and wine, salt pork and lentils, dried fish, cheeses, seven live cows, and three pigs. There were also sugar and figs, raisins and onions, mustard and rice, honey and flour. Fresh water had been loaded. The quality and quantity of rations were to be good by 16th century seamen's standards.

For trade, the ships carried bells, knives, silk and, cloth clothing. Finally for the wayward seaman, leg irons and manacles. No detail of necessity had been overlooked. With such good preparation, the crew was undoubtedly looking forward to a good voyage to a tropical paradise that promised riches for all.

Magellan had taken command of the 110 ton flagship Trinidad with a crew of 62 men. The Spanish captains were Juan Cartagena, who commanded the 120 ton San Antonio with a crew of 56, Gaspar de Quesada, head of the 90 ton Concepcion with a crew of 43, and Luiz de Mendoza over the 85 ton Victoria with a crew of 44. The aging Portuguese Juan Serrano captained the diminutive Santiago, a 75 ton craft with a crew of 31.

From the start, Captain General Magellan was faced with two problems; keeping the fleet of five ships with different sailing characteristics together and maintaining his authority over three Spanish captains. He accomplished both objectives by requiring the fleet to remain close behind the flagship Trinidad, and follow her lead in setting sails. A series of lantern signals from the Trinidad passed vital information to the other ships and were to be answered in kind by the rest of the fleet. Every evening, each ship had to approach the flagship and the captain of that ship had to greet the flagship with the words, "God save you, Sir Captain General and Master and good ship's company," thus acknowledging the authority of the Captain General and forcing the fleet to remain together. Only when this exercise was completed did they receive their daily orders.

Although all of the captains had sworn loyalty to Captain General Magellan, Pigafetta noted that they "hated him exceedingly." Certainly Magellan had the authority given him by King Charles to whom he had the utmost loyalty. His loyalty was returned with confidence. The first leg of the voyage to Tenerife in the Canary Islands off the coast of North Africa lasted several days. It was a time of shakedown and familiarization with the characteristics of the ships. In the harbor of Tenerife, a fast dispatch boat came bearing a letter for Magellan from his father in law, informing him that the three Spanish captains planned to take over the expedition and kill Magellan if he resisted. Magellan sent a return letter saying he would try to win the support of these men. We can only speculate on the Captain General's motives in ignoring this warning.

A meeting of the captains was held in the Trinidad's cabin. Apparently, captain Cartagena was planning to provoke a confrontation, giving him an excuse to kill the Captain General. He assumed command of the meeting immediately, issuing orders and ignoring Magellan's authority. Outnumbered, but calm and calculating, the Captain General acquiesced to all of Cartagena's bravado, thus postponing until he could more readily control the situation. Magellan had agreed to follow the course originally set by Cartagena. When the fleet sailed from Tenerife, they headed southwest as agreed.

At daybreak on October 5, the Trinidad suddenly changed course and headed south. Cartagena protested vigorously from the San Antonio, but had no choice but to follow or desert. With the four ships obediently following the Trinidad, Magellan had reestablished his authority and Cartagena realized that he had been cunningly tricked. Also by steering in a more southerly direction, Magellan avoided a fleet of Portuguese caravels launched in an attempt to stop the Spanish expedition. The course remained a mystery to the Spanish captains. The fleet sailed between the Cape Verde Islands and the west coast of Africa. On October 8, a terrible storm hit the fleet, rolling the ships, threatening to capsize them, and covering the decks with crashing waves. The storm later passed, leaving an eerie display of St. Elmo's fire in its wake.

This was followed by twenty days of doldrums. As the ships lay becalmed in the oppressive equatorial heat, with only Magellan knowing the reason, the crews were easily stirred by the undermining efforts of Cartagena. The ships pitched and rolled with the ocean swells, pulled along by the currents, but no wind filled the still, heavy sails.

Magellan remained serene while the crews grumbled, with little to do except seek shade under the motionless sails and wonder why their course had taken them out of their way into the horse latitudes.

During this time, the ships' captains could row back and forth between themselves at will in the ships' small boats. This would have made it easy for the Spanish captains to confer and foment mutiny. Heat began to take its toll on the meat supplies and the seams of the ships and barrels. The decks became frying pans and the holds became ovens. Finally, on the 21st day after the storm, the wind filled the sails and the ships were able to steer westward toward Brazil. On November 29, 1519, the fleet crossed the equator.

Now Cartagena began open defiance. He no longer gave the customary salute from the deck of the San Antonio. Magellan took no action, while the crews wondered what would come of this affront and challenge of the Captain General's authority. Morale was low as the seemingly erroneous course to the south that had lost two months' time was contemplated.

Then Captain Mendoza reported from the Victoria that two crewmembers had been caught in the act of sodomy, a common offense usually punished by flogging, but capable of drawing the death penalty. Knowing he faced a showdown with Cartagena, Magellan called for a court martial, and, for the first time at sea, called the four captains together. The seamen were sentenced to death, with the sentence to be carried out upon landing in Brazil. After the court martial, Magellan and Serrano were again facing the three Spanish captains in the cabin of the Trinidad. This was a replay of Tenerife with Cartagena trying to provoke Magellan into making an error.

Cartagena again began to berate the Captain General, who again remained calm and meek. Encouraged, Cartagena finally announced that he would no longer obey the foolish commands. Magellan was prepared. This was open mutiny in front of witnesses. He seized Cartagena with a burst of such sudden ferocity that the man was overwhelmed. The cabin quickly filled with men holding swords and daggers. Mendoza and Quesada, suddenly helpless, watched as Cartagena was clamped in chains. Magellan was definitely in command. Hoping to smooth over the wounds of the remaining Spanish captains, Magellan wisely agreed to release the Spanish nobleman to Mendoza's custody and appointed another Spaniard, Antonio de Coca as captain of the San Antonio. As the fleet continued on, Magellan enjoyed full command and the renewed respect of his men.

By December, the fleet was sailing near Brazil. These were Portuguese waters and the fleet stayed far from the coast with the gun crews ready. They were now in dire need of fresh fruits and vegetables after their three months at sea, and the antsy crew members were anxious to land for reprovisioning. Magellan ignored their eagerness and steered south until he was certain he was past any Portuguese settlements. On December 12, the fleet sailed into the beautiful natural harbor now known as Rio de Janeiro. Here the interests of the chronicler, Pigafetta, were aroused. He wrote lengthy descriptions of the natives and the flora and fauna during his first contact with the new world. He told about the natives believing that the small boats launched and gathering around the ships were children of the ships suckling from their mother.

As soon as the seamen discovered they could purchase one or two Indian maidens for the hatchet or knife, the Captain General faced a complete breakdown of discipline. After the long months at sea, here was a warm paradise. Even the constant threat of discovery by a Portuguese fleet could not maintain the gunners at their posts. One of Magellan's most trusted officers deserted the Trinidad, returning in the custody of a squad of marines. Further, the newly appointed Captain de Coca took it upon himself to release Cartagena, and Magellan demoted him for this act of insubordination. Short of men with leadership abilities, he placed his loyal cousin, Alvaro de Mesquita in command of the San Antonio.

On December 27, 1519 the fleet sailed from the delights of the Rio harbor. The natives had generously provided pineapples, sweet potatoes, sugar cane, geese, and chickens. During the two week respite, the sailors had also managed to take the time to fill their water casks. Temporarily satisfied, the men returned to the sea well fed, rested, and presumably smiling. A warm southerly breeze moved them down the South American coast.

Magellan was inwardly concerned. His ships' captains were untrustworthy, some clearly disloyal, others of marginal competency. Driven by his vision, but wary of the captains and crews, the Captain General set out in search of El Paso, the legendary strait through the New World.

The armada sailed south to 35° south latitude covering territory familiar to the fleet's pilots who had explored the area with John of Lisbon six years earlier. It was here at the broad estuary of the Rio de la Plata near present day Montevideo and Buenos Aires, that Magellan hoped to find El Paso. He sent the tiny Santiago, which was well suited to exploring tight places, into the passage. Convinced that he was close to his goal, he took the San Antonio and sailed off in search of the theoretical Terra Australis, the Southern Continent rumored to be in the area. Upon his return, he met with Juan Serrano of the Santiago. Adding what he had seen to the information provided by Serrano, Magellan realized that he was merely at the wide mouth of a river, and that only more of the South American continent lay ahead.

Now the character of the voyage began to change. El Paso was not where he planned, and Magellan faced the unknown. Unable to confide his fears to his captains, and unwilling to admit them to himself, he doggedly drove his fleet southward along the coast exploring every inlet. The weather began to turn sour as they pushed toward the Antarctic. Winter was approaching that part of the world, but Magellan had to find El Paso. They passed islands full of penguins, which Pigafetta duly recorded as strange geese, and seals. The landscape grew more bleak and somber. The ships were tossed by storms that brought fog, snow, ice, and frigid temperatures. The officers begged the Captain General to turn back, but he seemed to have become more obsessed, irrationally determined to reach the elusive El Paso.

Finally, at 49°south latitude, they found a bay with calm waters, edible fish and fowl, and a supply of fresh water. Feeling he dared go no farther, Magellan announced that they would build shelters, and winter here in the place named Port Julian. After six months of travel, the complement now faced a freezing winter in a barren and desolate land. This decision had been made without consultation, and was taken by the Spanish captains to be a final insult. The men again murmured against the Captain General who had brought them to such an ill state. Settled in with an easily swayed crew and mutinous captains, Magellan faced the problem of the missed passage. Had he continued but a few hundred miles south, he would have been vindicated by finding the El Paso, and possibly saved himself many miseries and much time.

On Palm Sunday, April 1, the Captain General invited Cartagena, Mendoza and Quesada, as well as the newly appointed Mesquita, for breakfast in the cabin of the Trinidad. Only Mesquita appeared. There was no mistaking what was happening, but Magellan was not shaken, nor was he forced into immediate action.

At midnight, Captain Quesada took 30 men from the Concepcion and rowed with muffled oars to the San Antonio. They boarded the vessel, seized the sleeping Mesquita and clamped him in chains. Unfortunately, the master awoke at a most inopportune time, and, in a vain attempt to get Quesada to leave, he was stabbed six times and was mortally wounded by Quesada. Crewmen loyal to Magellan were chained. A young Basque, one of a group recruited to round out the fleet's roster, was placed on the San Antonio. His name was Juan Sebastian Del Cano. This mutineer was destined to captain the expedition home. Although Magellan was to be remembered, Del Cano was to be the first to circumnavigate the globe in a single voyage.

On the Victoria, Captain Mendoza released Cartagena, who rowed over to take command of the Conception. Now the big San Antonio, the Conception, and the Victoria were in the hands of the mutineers. When Magellan awoke the next morning, he faced them with only the Trinidad and the little Santiago.

A boat from the San Antonio delivered an invitation to Magellan to come to Quesada to discuss terms. Magellan was cold and calculating with a quick and ruthless plan. First, he seized the messengers, gaining a boat and depriving the mutineers of some of their fighting men. Then he wrote a letter to Mendoza, and sent it with five men carrying concealed weapons to the Victoria. The crew of the Victoria certainly would not expect an attack from a mere five men in a small boat. The message was delivered to Captain Mendoza, along with a quick, fatal dagger thrust to the throat. Simultaneously, 15 well armed men boarded the Victoria from the Trinidad's other boat. Overwhelmed by the sudden reversal of fortunes, the Victoria's crew acquiesced to the command of Duarte Barbosa and hoisted anchor, joining Magellan's other two ships.

Panicked at the sight of the three ships blocking the harbor, Quesada made a bungled attempt to get the San Antonio past the other ships. With their support fading under the rapidly changing balance of this ill-fated mutiny, both Quesada and Cartagena surrendered.

Over 40 men had participated in the mutiny. A quick court martial was held. Mendoza, already dead, was hanged, drawn, and quartered. Quesada was beheaded and quartered. Cartagena and an especially troublesome priest were left marooned in this desolate land when the fleet sailed. Others, including Del Cano, were placed in a chain gang and served hard labor until the stay in Port Julian was over.

After the grisly court marshal and sentencing, the men settled into a gloomy cold, keeping themselves provisioned by hunting and fishing. They built shelters and made warm clothes of skins. They careened the ships on their sides for caulking, cleaning, and patching. The only diversion was the appearance of giants. These giant men were seen by several expeditions in the area as time went by. Pigafetta took note of these huge men. They ran nearly naked and seemed unaffected by the cold. Eventually these natives befriended the Europeans, giving the men a close look at the men whose large feet gave the area its name, Patagonia. It was reported that one of these men could bellow like a bull, swallow a bucket of water in a gulp, and stuff half a hamper of ship's biscuits into his mouth. Relations with the natives were good until Magellan, under orders to bring back specimens of flora and fauna, decided to capture two of them. He loaded them down with gifts until they could carry no more and added leg irons as the final present. One escaped to tell the story of Magellan's treachery to the other natives. From then on, the natives were at war with the Europeans. One crewman was killed by a poisoned arrow, giving the travelers their first death outside of the havoc they were creating against each other.

While the ships were being overhauled, a complete accounting was made of the ship's provisions. An appalling shortage was uncovered. Portuguese sabotage had been well at work on the docks of Spain, and falsified consignments had completely deceived the ships' pursers. Magellan hid their plight from his captains in his usual one-man-show manner. He immediately set out to pack the barrels with smoked and salted fish and fowl, but there was a severe shortage of fresh fruits and vegetables.

With the first hint of spring, the Captain General sent Serrano out in the Santiago to scout for El Paso further south. Immediately upon leaving the sheltered harbor of San Julian, the Santiago found herself in the midst of the stormy winter gales of the southern latitudes. For sixteen days they struggled south covering only about 60 miles and discovering the harbor of Santa Cruz. In attempting to return, the Santiago was struck by another storm. A strong wave ripped the rudder away, and Serrano managed, with careful use of the shredded sails to reach a sandbar where the men jumped to safety. One man was swept away and drowned, and the other forlorn survivors stood and watched as their ship was pounded and swept away by the waves.

The 37 remaining men managed to find some planking along the shore, and two volunteers crossed the bay on a makeshift raft to make their way overland for 11 days, without food or weapons, and only snow to drink. Magellan was able to send a rescue party by land and return the stranded survivors.

On August 24, 1520, the four remaining ships left Port San Julian and sailed to Santa Cruz where the wreckage of the Santiago was discovered and stripped. The crew of the Santiago was apportioned among the remaining ships. Magellan placed Duarte Barbosa in command of Mendoza's Victoria, Serrano took Quesada's Concepcion and Mesquita, who had been placed in chains by the mutineers, was restored to command of the San Antonio.

After a year, the fleet had still not found El Paso. Thus far, the loss of life had been minimal. There were the two men presumably executed for sodomy in Rio Harbor, two executed captains, one killed by Patagonians, and one lost in the wreck of the Santiago. There were also two that had been marooned and probably did not survive for long.

The captains wanted to turn back, and Magellan finally agreed to do so if El Paso was not found by the 75° south latitude. It was now obvious to all that Magellan did not know where the elusive pass was, or if it existed. Unknown to them all, it lay only two sailing days south of Santa Cruz. Bowing to the present pressure of the severe weather, Magellan stopped and wintered another two months at Santa Cruz.

On October 18, 1520, the ships weighed anchor and continued south through storms to a place which they called the Cape of Eleven Thousand Virgins. The Captain General called another meeting of his captains. To his dismay, he found that even the loyal Barbosa, Serrano, and Mesquita were opposed to further exploration. These men were not far removed from the idea of a flat earth whose treacherous edge one could simply sail off and into the abyss. The time at sea was wearing even on the stoutest of heart. Undaunted, Magellan ordered the San Antonio and the Concepcion to explore the bay. The reluctant captains obeyed. Then the Captain General watched helplessly as the two searching vessels were swept by a gale to near certain destruction against the rock lined shore. Separated, the Victoria and the Trinidad fought the gale. Their decks were pounded with the waves, and they continued to struggle for survival for two interminable days, until the storm at last abated.

When the waters and the skies cleared, the Trinidad was in sight of the Victoria. As soon as possible, they entered the bay to look for the two wrecked ships. Instead, the surprised lookouts saw first one sail, and then another. Both ships were safe. To the further astonishment of the gale weary crews, the two wayward ships had their flags out and their guns firing. When Mesquita came alongside the Trinidad, he excitedly reported that he had found El Paso, a strait where the waters were salt and they ran both directions. No longer would they have to struggle south toward even colder lands. The men danced and cheered at this apparent miracle.

The other captains wanted to note the location on their charts and turn east. The fleet was short of provisions, and the area to the west was unknown. Magellan declared forcefully that they would go forward, believing that the Spice Islands were a short distance from the strait. The fleet sailed together into the Strait of Magellan. At a fork in the waters, the San Antonio and Concepcion went one way, and the Trinidad and Victoria went another.

It was a boat from the Trinidad that finally found the broad, beautiful Pacific Ocean. Meanwhile the San Antonio was separated from the Concepcion and taken over by mutineers. The pilots had clamped Mesquita in irons, and returned to Spain, arriving in May of 1521, with wild stories of Magellan's cruelty and mistreatment of the Spanish noblemen. The desertion of the San Antonio, while mutinous and criminal, may still have been a more rational act than driving the exhausted crewmen into the unknown ocean with inadequate provisions. History may judge.

The Captain General searched in vain up and down the strait, believing the San Antonio to be wrecked. A resident astrologer gave them a correct answer: that the San Antonio had been taken by mutiny. Immediately, Magellan ordered an inventory of the remaining ships. The 120 ton San Antonio had been the fleet's largest ship. She was heavily armed, and had carried much of the vital supplies, and held a goodly portion of the remaining crew.

In a rare move, Magellan confided in captains Barbosa and Serrano. They did not want to return failures after the San Antonio had beaten them to Spain. With desperation and fear of shame in their minds, they now felt that they had to move forward. Once into the Western Sea where no Christian man had sailed before, the fleet turned north. They were anxious to enter more tolerable climates. The water casks and larders had been filled as much as possible. At 42° south latitude, they still had not left the coast of present day Chile. They persevered onward, ever watching for the coast of Asia, which they hoped jutted close to this monstrous American continent. They continued northward in the strange seas until 32° south latitude, and then turned west into a frightening emptiness. Stretching endlessly before them lay the broad Pacific Ocean. Soon starving men in tiny ships would face the sight of endless unbroken miles of water and sky. It was almost the final blow to the soul sick men.

The date was December 18, 1520.

For three months and 20 days they plodded this vast and barren sea with a warm sun and a light breeze. Even the welcome change in the weather could not hide the desperate state of their dwindling rations. The poorly smoked meat putrefied. The water turned yellow. They ate the biscuits until this life saving staple was depleted. They ate crumbs, worm eaten and yellow with mouse urine. Crews caught mice and rats to make meager meals. Still they saw nothing but the sun and gaunt sunken faces of their comrades in unspeakable misery.

In time a mouse brought half a ducat, as recorded by the longsuffering historian Pigafetta. Nineteen men died from scurvy, their gums so swollen, they could not eat. The captured giant, and the Indian crewman that had been picked up in Brazil, also died. Twenty-five or thirty became so ill that they were unable to help man the decks. Finally, they were reduced to soaking the hides, which were wrapped around the masts, until they were soft enough to put in a pot over a fire. This they ate with the available sawdust. A storm would have been the end of the weakened crew. Pigafetta wrote, "If God had not given us fine weather, all would have perished of hunger in this very vast sea." He went on to explain that all were certain such a voyage would never again be attempted.

During the crossing, they had seen two small islands. Out of mice and rats, they landed on one of these islands to trap a few birds and hunt for what eggs they could find. They were also able to supplement their brackish water supply by catching rain from a short squall. On March 5, 1521, with absolutely nothing left on which to survive, land was sighted. As they approached the parcel of land, now called Guam, the natives came out in their canoes. With little knowledge of private property, and quickly perceiving the weakened condition of the Europeans, they began to loot everything that was not nailed down, pushing the feeble crewmen aside with open disdain. The situation got so out of hand that Magellan ordered a crossbow volley.

The locals and the Europeans were now at war. The next day, the three ships moved in and fired guns into the village, then launched a contingent of armed men to the island. The aggressive approach had worked in Magellan's favor, and the natives did not appear while the men filled their water casks and looted the village of rice, yams, coconuts, chickens, pigs, and bananas, welcomed by Pigefetta as "long figs". The fleet quickly retreated from the island while the native dugouts swarmed around them with shouts and taunts.

With the provisions gained here, and some trading on a nearby island, the sick began their slow recovery. The fleet continued west and reached the Philippines on March 16, 1521. Surely now they were close to the Spice Islands and loading up with the cargo that would make them rich. Already they had done what had never been done before. They had found the legendary El Paso, and crossed the Great South Sea.

The Captain General anchored the three ships in a quiet bay near a wooded island in the Leyte Gulf. After so much suffering, they were again in a warm paradise. The sick were brought ashore and a shelter built for them. A stockade was erected. The ships were careened, and repairs were made. Fish, food, oranges, and coconuts quickly brought the crew back to full strength. After ten days, they continued their exploration. At one point, Enrique, Magellan's slave, hailed a canoe and was answered in his native Malayan tongue. Now it was proven. They had reached the East by sailing west.

Magellan felt he would be the future governor of the Pacific for Spain. The fleet tarried for a while in the Philippines, his planned future domain, where they traded iron for gold. On April 7, 1521, the three ships entered Cebu Harbor. There they found a well-developed village, a Chinese junk, and a fairly prosperous Rajah. The Europeans presented themselves as generous friends, but unconquerable enemies. The Rajah was easily convinced, having heard of the excesses of the Portuguese in times past. The Rajah acquiesced to the peaceful trade. In this seemingly peaceful time, with the loyalty of his men and the cooperation of the local leader, however, the chain of events would begin that would stop Magellan's dreams.

Perhaps because of the survival of such a remarkable voyage, Magellan became obsessed with religious zeal and Spanish invincibility. While his crew traded scarce and valuable iron and brass for plentiful gold, spices and sexual favor, Magellan tried to instill Christian virtue in his men and pass it on to the natives. Impressed by the ceremonies of the Europeans, even the Rajah declared his willingness to be baptized. Amid much pomp and ceremony, the islanders became a "Christian" community. Yet there was a holdout on the nearby island of Mactan. The Captain General had always been fond of demonstrating Spanish power for the natives by having his men thrust daggers at another soldier in armor. This seemed to impress him as much as the natives, for he decided to take a contingent of 60 men and show how easily they could out fight the 1,500 warriors arrayed against him. The Rajah offered to even the odds with his own men, but Magellan seemed determined to show the superiority of a few Spaniards armed with muskets and swords and protected by armor. This certainly did not seem like the cool and calculating Magellan, but it did fit the unswervingly determined Magellan.

On the night of April 26, the Captain General took his men, mostly untrained volunteers to Mactan in longboats. After first sending an envoy with terms which were denied by the Mactan ruler, he waited until dawn. The first difficulty encountered was a coral reef barrier that prevented the longboats with their protective small cannon, from taking the troops into the island. Eleven men were left to man the boats, and 49 men in full armor struggled waist deep in water for two bowshots distance to get to the battleground. They arrived on the island exhausted from the effort. Pigafetta accompanied this group. When they had reached the island, they found an effective barrier of trenches, and a formidable enemy that was not easily cowed by their blustery appearances. Magellan had not taken his trained fighting men, nor set up artillery cover from the fleet. The men who did reach the island used their weapons ineffectively. The natives quickly realized that the sophisticated crossbows and arquebuses were noisy, but nearly useless. They also discovered that even with armor, certain portions of the Spaniards bodies were vulnerable. Singling out the Captain General, their spears and arrows homed in on the faces and legs.

The ill prepared forces lacked commitment to this wild venture. When they were put in danger by being cut off from the sea, they broke and ran. The Captain General, wounded in the face and on the arm, had only the loyal Pigafetta, his slave Enrique, and two others to stand by him as he tried an orderly retreat. The natives fell on him en masse.

Pigafetta, himself sorely wounded, managed with the others to retreat to the boats, which were already pulling away. Eight others had been killed in this hapless battle and several drowned. The Captain General was mortally wounded and died soon on the island. With Magellan's death, a leadership void opened up in the fleet. It had been his dream, his driving force, his personality and strength that had brought them across two oceans, and the void seemed a deep one.

Barbosa and the aging Serrano were in charge, and hit by immediate indecision. The newly converted Rajah realized that the Spaniards were far from invincible, and took matters into his more capable hands. He sent word to the Mactan natives, asking if they would return the body of Magellan. They refused. Enrique, wounded and grieving over the loss of his beloved master was rousted out by Barbosa. The new leader reminded Enrique that he was a slave and ordered him to go ashore as an interpreter. The brutality of Barbosa's order and the loss of respect for the new captain by the Rajah made a deadly combination.

Enrique concocted a plot with the Rajah designed to get the ships and weapons of the Spaniards. They invited Barbosa to a farewell dinner and to receive a valuable gift for King Charles. Naively, he and 23 ranking officers went to this dinner. Two returned to the ships when they became suspicious. No sooner had they finished voicing their suspicions to the men on the ships when shouts and cries came from the shore. The ships quickly cut anchor and moved in, firing their cannons at the house. Suddenly Serrano appeared on the beach bound with ropes. He shouted that the others were dead and begged his shipmates to ransom him. Instead of trying to save him, the fleet backed off. Pathetically, Serrano wept and yelled to the pilot Carvalho, now in command, that he would ask Carvalho to account for his soul on Judgment Day.

This senseless war and slaughter again changed the complexion of this remarkable odyssey. There were only 108 survivors, and they were leaderless and lost. They elected the veteran pilot Carvalho as the new Captain General. Not only did he lack the piloting skills he needed, he was also bereft of leadership ability and moral fiber. Without sufficient crew to man three ships, they transferred the crew and cargo from the Conception and set it afire. It was May of 1521, and to some of the watching sailors, the Conception had been their home for two years, carrying them halfway around the world.

For months, Carvalho led the fleet in a piratical wandering in the South China seas, robbing and sinking Arab trading ships, Malay prahus, and Chinese junks. He kept a harem of captured women in the cabin of the Trinidad, and was finally deposed by the other men, who became ashamed of the degeneration that had taken place. They had been hired for a journey financed and commissioned by King Charles, and they could not continue as lowly thieves. They then elected Gonzalo de

Espinoza, captain of the Victoria as Captain General. He moved his command to the Trinidad and appointed Juan Sebastian Del Cano as captain of the Victoria. Del Cano, whose earlier death sentence as one of the mutineers in San Julian had been commuted by Magellan, was now second in command.

Espinoza, like a lost traveler stopping at filling stations, methodically worked his way toward the Spice Islands, inquiring at each port. He restored discipline and stopped the piracy. On November 8, 1521, the two ships entered Tidore Harbor in the Spice Islands. It had taken them twice as long to wander to their goal as it had to cross the Pacific. Pigefetta dutifully recorded each island and its peoples.

Now they could start serious trading. The European metals were as valuable here as spices and gold were in Europe. Espinoza built a fortress to establish a Spanish foothold, and a warehouse to store unused copper, iron, and other trade goods. The ship's holds were cleared. The ships were then loaded with spices for the return. At the last minute, the faithful flagship Trinidad, heavily loaded for the trip home, literally burst her seams and began taking on water. Winds from the seasonal monsoons were necessary for a quick voyage home. In order not to miss the monsoons, it was decided that Del Cano would take the Victoria home via the Cape of Good Hope. With a crew of 47 Europeans, 13 Moluccans, and a cargo of 26 tons of Cloves, the Victoria weighed anchor in late December. The Trinidad wasn't made seaworthy until April of 1522. By this time, the monsoons had passed and Espinoza tried to take her east across the Pacific. The Trinidad struggled unsuccessfully against the formidable headwinds and wandered the Pacific until her crew began to perish. She returned to the Spice Islands where she fell into the hands of the Portuguese. Her crew was held in captivity for four years. At the end of that time, only Espinoza and four others were alive and released to return to Spain.

The return voyage of the Victoria was as arduous as the one which carried them to the East. Avoiding the Portuguese controlled areas of Asia and Africa, Del Cano took the lengthy course across the Indian Ocean. Twenty one sailors died in this crossing as the tropical heat rotted the meat and provisions picked up in the islands. Scurvy was rampant.

In desperation, Del Cano stopped at the Portuguese harbor of Santiago in the Cape Verde Islands. The Spanish sailors claimed to have been on their way back from America, and bought food. Here Pigafetta, onboard and still recording events, was surprised to find that it was Thursday, July 10. He had kept a detailed diary, and his diary said it was Wednesday, July 9. The dateline had been crossed. The circumnavigators had observed a strange new phenomenon. A new problem brought an end to the speculation on the dates. The Portuguese caught on to the ruse, and arrested a 12 man shore party. Quickly Del Cano set sail for Spain. On September 8, 1522, the battered little Victoria returned carrying 18 survivors of the original crews. This completed the longest voyage in the recorded history of man, around the world in two years, 11 months and 17 days. Final proof of the size and shape of our world was now known.

The single cargo brought sufficient income to pay for the cost of the entire expedition, a boon to the investors, who had written off the expedition as a total loss. As was customary, the sailors received little, and the families of the dead received nothing.

Magellan, whose driving vision and determination had started it all, had not survived. His second child was stillborn and his wife and first child did not live to see the return of the Victoria. The world had been encompassed, but at a heavy price. Del Cano, the first circumnavigator, became so by accident.

Monetary gain and national rivalry had motivated the first voyage and record. War would motivate the second.

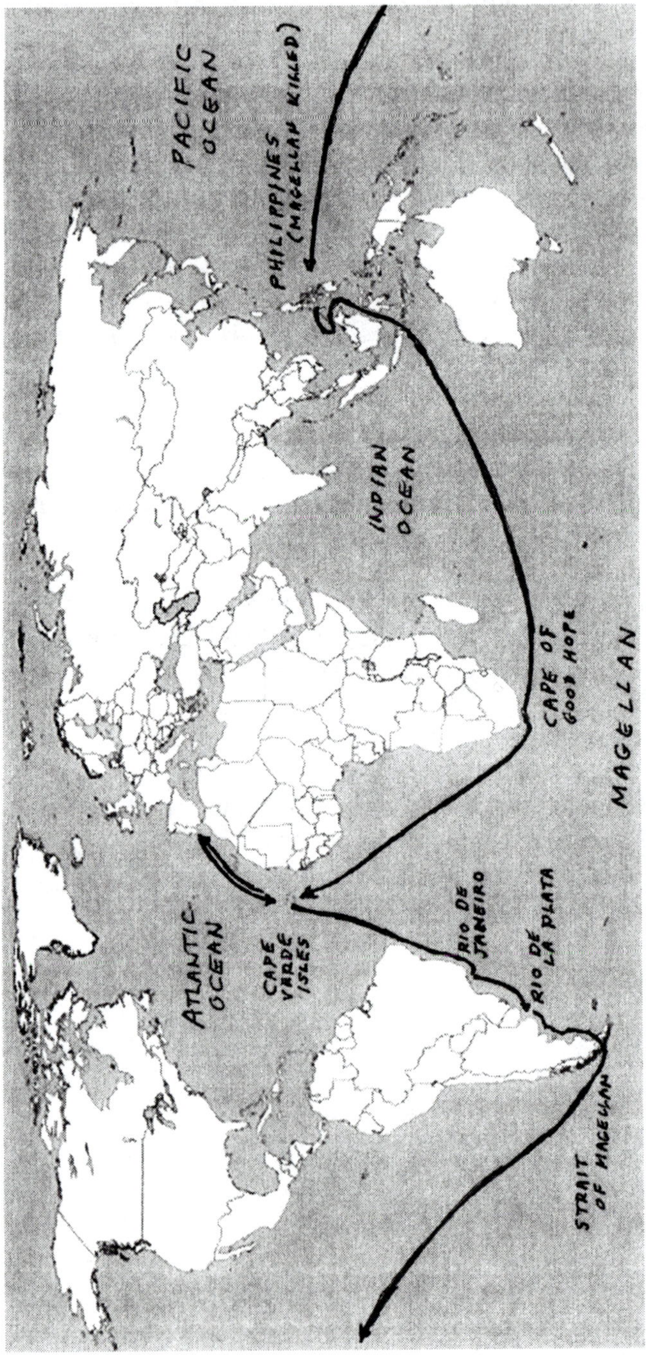
PACIFIC OCEAN
PHILIPPINES (MAGELLAN KILLED)
INDIAN OCEAN
CAPE OF GOOD HOPE
MAGELLAN
ATLANTIC OCEAN
CAPE VARDE ISLES
RIO DE JANEIRO
RIO DE LA PLATA
STRAIT OF MAGELLAN

CHAPTER 2 - DRAKE

It appeared that the prophecy of Pigafetta and his fellow sailors would prove correct. For years, it seemed that such a voyage would indeed never be made again. [1]

Some attempts at circumnavigation were made unsuccessfully. The trip was both rigorous and unnecessary, since it was now known that it was faster to go around Africa. The Spice Islands, won by Magellan at a dear price, were sold by the Spanish king to Portugal for a nominal sum only a short time later. Thus Portugal regained her control of the East and a foothold in Brazil.

Spain, on the other hand, found wealth in the New World, in the form of gold and silver. The conquistadores brutally explored these lands. The mapmakers, fresh from revising their estimates of the earth's size, were busy filling in the details. From Panama, Spain was able to establish settlements along the western coast of South America. These provided a solid basis for systematic looting of the New World and religious missionary work. The political ascendancy of Spain quickly eclipsed Portugal, and soon her power extended to dominate Europe. France remained consumed with her own internal struggles, and England

---

[1] According to one source, there were two circumnavigations started prior to Drake and after Magellan. Two Spaniards named Grijalva and Alvaradi are listed as circumnavigating in 1537. A man named Mendana later completed a circumnavigation of sorts. Mendana had discovered the Solomons in 1568. After later moving to Peru, he left to settle in the Solomons in 1595 at the age of 53. Upon reaching the Solomons, he had completed a circumnavigation.

smarted from Spain's growing rule of the seas.

Religion had not departed from the political scene. The struggle shifted from Moslems and Catholics to Protestants and Catholics. Pastor Fletcher, who sailed with Drake, summed up a prevailing attitude when he said, "The Africans made a god of the sun, which, since they knew no better, was reasonable, but the Roman Catholics, though they knew of God, preferred to worship images."

The Spanish Inquisition was an attempt to retain the power of the Catholic Church by cruel force. Many an English seaman fell victim to the raging Inquisition, which degenerated into incredible excesses of torture and inhumanity in the name of religion. Anyone suspected of being a Protestant ran the risk of torture, forced confession, and death.

English merchants and seamen suffered, then retaliated. Piracy took on new meaning as the English royalty punished raids on Spanish shipping with an official reprimand and a wink. In 1545, an English captain named Reneger made himself a future life of ease by capturing the Spanish treasure ship San Salvadore with her cargo of sugar, hides, and gold.

The hope of such affluence helped motivate Francis Drake. The son of an impoverished preacher, Drake had gone to sea at an early age. He came up through the ranks, and became one of England's finest navigators. While sailing under another exemplary English captain, John Hawkins, he barely escaped a treacherous sneak attack by a Spanish fleet. Most of Hawkin's men were massacred and only two English ships escaped to sea. One of them, the 50 ton Judith, was commanded by Drake. England and especially Drake were incensed by the attack. Drake dedicated himself to piracy and puritanical religious zeal. The West Indies and Panama yielded up their treasures to the energetic Drake, and he became a man of wealth and pretension. It gave him the wherewithal to invest in a bold mission.

Like Magellan's mission, this was not intended to be a circumnavigation, but was meant to return by the same route. Also, where Magellan had an entourage of Spanish nobility, Drake was to have a troublesome contingent of nobility who showed disdain for the common seamen. Unlike Magellan, Drake was allowed to carefully choose many of the seamen. The English had not established themselves in distant seas, and most of the men signed on without knowledge of their destination.

Drake was established as a commander, but his lowly birth gave him questionable authority over the gentlemen who would accompany him. One of these men, Thomas Doughty, was a close friend and knew his way around the English court. He was influential in getting the backing necessary for the voyage. The Queen provided her blessing. Preparations for the voyage were carried out in utmost secrecy to avoid tipping off the Spanish.

One hundred and sixty-four men sailed from Plymouth under Drake's command using five ships. The flagship was the 120 ton Pelican, which would later be renamed the Golden Hind. The queen contributed the 90 ton Elizabeth, commanded by John Wynter, and an investor contributed the 30 ton Marigold, commanded by John Thomas. John Chester was over the 50 ton supply ship Swan. The 15 ton bark Christopher rounded out the fleet as a shuttle vessel. Three pinnaces were lashed to the decks for later assembly. Although the fleet totaled only 295 tons, considerably less than Magellan's 480 tons, and was manned by less than two thirds the number of crewmen, more men were destined to return safely at the end of the excursion.

The Pelican was a good fighting ship. Magellan's Victoria relied on her big square mainsail. The more modern Pelican carried topgallant sails fore and aft. She carried fourteen long range cannon firing nine pound shot and swivel mounted falconets on deck. The men were armed with pistols and arquebuses, bows, arrows, pikes, fire bombs, and swords.

Perhaps the most unusual equipment was the trappings of wealth and luxury. Drake's cabin was stocked with carved chairs and tables, silks, oriental rugs, fine clothes, and silver utensils. The 38 year old admiral obviously intended to travel ostentatiously, having at least the moneyed appearance of gentry, if denied their bloodlines.

Packed with spare sails, timber, spars, ropes, blocks, casks of biscuits, beans, lentils, salt pork, beer, cheese and honey, and live penned chickens and pigs, the entourage set sail on November 15, 1577. On November 27, they limped back to Plymouth harbor so badly battered by storms that they could not continue without repairs. This was an inauspicious beginning for the superstitious sailors, but did little to deter Drake from his goal of raiding the Spanish in the great South Seas.

On December 13, 1577, the sky finally cleared, and the fleet sailed from Plymouth. As soon as the fleet was at sea, Drake set Mogador, just south of Gibraltar on the West African coast, as a rendezvous point. Morning and evening aboard the Pelican the men gathered for worship, led by the Puritan preacher Francis Fletcher, or by Drake himself.

On December 27, they entered the harbor at Mogador. They met the long robed Berber tribesmen, whom they invited to a banquet at the captain's table on the Pelican. They offered gifts, and established trading relations to be consummated the next day. When the Berbers appeared the next morning, Drake sent a boat ashore to continue the friendly barter. Instead, several armed men leapt on to the first sailor on shore, a man named John Fry, and carried him off on horseback. The other men in the boat stood by with oars to fend off the attackers, but, as the Berbers rode inland on horses and camels, they could only row back to the ships. Any attempt to chase them now would have been futile and extremely dangerous. Drake weighed anchor and headed southwest along the barren, hot coast of Africa. The expedition has suffered its first loss. Actually, John Fry was courteously treated and eventually put on an English ship in the Mediterranean.

While at Mogador, they had assembled one of the pinnaces. The pinnace would be useful in taking soundings close to land, for carrying messages, for quick privateering raids, and for boarding Spanish ships. She could be carried on board the larger ships, disassembled and built when and where she was most needed. All of the early circumnavigators carried such boats.

The swift little pinnace was used by Drake on January 7, 1578, to capture three Spanish fishing vessels. Drake forced the fishermen to accompany him as far as Cape Blanc in West Africa. In the harbor there, Drake pillaged the Portuguese ships at anchor, taking all of the available food and any useful charts and maps. By this time, a faction of the crew, including Wynter and Doughty, began to be concerned about the overt piracy, while the determined Drake drilled the men in combat.

Now they abandoned the little bark Christopher and took one of the larger Spanish vessels, which was perhaps 30 or 40 tons.

Forcing a Portuguese Caravel to accompany them, the motley fleet set out for the Portuguese islands of Cape Verde, where several of Del

Cano's crew had been captured near the end of their journey. Once again, Drake hopped into the pinnace in pursuit of a prize, this time two Portuguese ships. He seemed contented to take both Spanish and Portuguese ships, showing his disdain for all Catholic countries. Drake overtook the first ship quickly and the captain offered no resistance. However, he had to turn back from his second prey when they came too close to the guns of the local fort.

The fleet, along with the newly captured caravel, headed to sea. The prize turned out to be just what the expedition needed. It was the 100 ton Santa Maria of Lisbon, quickly renamed the Mary by the Protestant Drake. She was heavily laden with supplies for settlers in the New World, and such goods were suited to the needs of a long sailing expedition. She carried canvas and warm clothing, nails and tools, and 150 caskets of wine. Most importantly, Drake found aboard a Portuguese pilot familiar with the South Atlantic. His name was Nino de Silva and he and Drake would become good friends before the journey was complete.

Upon reaching the most westerly of the Cape Verde Islands, Drake released the Portuguese caravel and put the crew of the Mary into the pinnace. He then placed his friend Doughty in command of the Mary. To the captured pilot de Silva, he promised anything except his freedom.

With a fleet consisting of six ships, the Pelican, Elizabeth, Swan, Marigold, the new Christopher, and the captured Mary, Drake set out to cross the Atlantic on February 2. The fleet moved into the area of the doldrums, one of the most dangerous threats to a sailing expedition. They were becalmed for three weeks and were 54 days without sight of land in the equatorial Atlantic.

Rations were cut in the face of uncertainty about reaching land. This tedious and devastating time affected Drake's crews as much as it had Magellan's. Now Thomas Doughty showed that the qualities that made him invaluable in financing and selling the expedition to the crown were not valuable in a long sea voyage. Doughty began to gather his own factious group and to be contentious, making small but open slurs at Drake's authority. The rift between the gentlemen and the commoners on the long journey had reached a taut standoff. Drake transferred command to the Mary, and put Doughty over the Pelican to separate the two factions. The jibes and pranks continued at Drake's expense. Finally, Drake was forced to banish Doughty to the supply ship Swan.

Here Doughty steadfastly continued his efforts to undermine Drake. When one of the officers suggested that Drake should deal with the mutineers as Magellan had, Doughty insisted that Drake did not have that kind of authority over gentlemen. The gulf between the seamen and the nobility had become dangerously wide.

The only interruption to the tedium and the infighting were the occasional visits of flying fish and violent thunderstorms. These proved helpful, by supplementing the food and water supply respectively. The captured pilot de Silva provided great service by guiding them across the Atlantic, far from the usual English seas. He also served posterity by recording the crossing. He kept the fleet away from Portuguese galleys, and brought them in sight of land on April 5.

Suddenly, the fleet was engulfed by a dense fog, the beginning of two weeks of terrible weather for which the region was famous. In the midst of fogs and storms, the new Christopher was lost. For days they searched for her, and then moved on to the Plate Estuary. Soon after their arrival, and much to the joy of Drake, the Christopher sailed in behind them. The ships continued up the estuary until they found fresh water. They refilled their water casks in three fathoms of fresh water, and added some freshly killed seals to their larders, finding their skins and blubber most useful.

No sooner had they set out to sea than a violent storm separated the Swan. They sought the Swan for only one day since the food situation was becoming critical. They had to seek a safe harbor where they could find fresh vegetables and meat. The last of the salt pork had rotted and the biscuits had become wormy.

Within a couple of weeks the fleet was again separated. The Pelican and Elizabeth were together, the Marigold and Christopher were together, and the Mary was by herself, while the Swan remained lost. It was Captain Thomas of the Marigold that found a good natural harbor, which was later named Port Desire. It was nearly 48° south latitude. From here, the Marigold and the Christopher sailed forth each day in search of the others. The first ship they found was the Swan, by now believed to be permanently lost. By the end of three days, they had located the other ships.

In Port Desire, two of the ships were beached and careened for recaulking. The Swan was emptied, stripped, then towed to the shore and burned. Drake felt this move consolidated his fleet and would therefore cut down on the time lost searching for separated ships. Now the crews were treated to the sight of the native Patagonians. They gave conflicting reports of the large men, each man having his own interpretation of "giant." There was a vast array of birds, eggs, and fresh meat for storing in the larders.

Drake transferred Captain Gregory to the command of the little Christopher and heard more and more of Doughty's insolence. When ordered to the Christopher, Doughty refused. Drake asserted his authority by ordering the cursing and screaming Doughty lashed to the mast. By nautical standards, the punishment was a mild one, but by gentlemen's standards it was intolerable.

As soon as they put to sea, the ships were again separated. Drake searched for his fleet, resolving that this would be the last time he would undertake such a hunt. He decided to reduce his fleet in size. The Spanish fishing bark Christopher and the Portuguese caravel Mary were chosen as dispensable. The Christopher was scuttled, and the Mary was broken up and used to build shelters in port San Julian. With the Swan, Christopher and Mary gone, the Pelican, Elizabeth, and the little Marigold became the consolidated fleet under Drake's command.

Drake reached port San Julian on June 19, 1578, six months and six days out of Plymouth. They found the gibbets on which Magellan's mutinous captains had been hanged. It had also taken Magellan approximately six and a half months to reach the same point from Seville.

Drake lost two men in a short, unexpected battle with the Patagonians, causing the need for a constant armed guard over their camp. In excavations, the seamen dug up the skeletal remains of those who had died in the 1520 mutiny.

Now Drake would be forced to deal with his own mutinous captain. Soon he would lead his men through the Strait to attack Spain in an undeclared war. Doughty's maneuverings were undermining and jeopardizing the mission. Drake impaneled a jury against his onetime friend, and acting as judge and prosecutor, ordered a death sentence. Captain Wynter offered to keep Doughty in chains and return him to England. Drake refused.

The troublesome Doughty mustered his dignity, and on July 2 Doughty and Drake took communion together. They went to Drake's cabin for a pleasant chat and enjoyed dining on dishes brought by servants. The men toasted the success of the voyage, and then rowed to a place on shore where the entire complement gathered for the execution. Drake and Doughty talked privately for a few moments, and then embraced. Doughty's last words were recorded as, "Strike clean and with care, for I have a short neck."

The execution marked another parallel with the voyage of a half century before. History would soon repeat in other ways. It was the middle of the Antarctic winter, and everything froze. Snow, ice, and freezing winds met every outside venture. Morale was low.

In August, Drake mustered the compliment together. In a masterful move of bluff and power, he offered to let anyone take the Marigold and leave for England, but warned them that they had better not run across him at sea. No one moved. He dismissed the captains and officers from their positions. When they protested, he made an eloquent speech about his direct authority from the Queen and restored them to their posts. He instilled in them the patriotic importance of their mission. The men were impressed, and with the recent execution of Drake's friend fresh in their minds, Drake could now face his goals firmly in command.

On August 17, with the weather improving, the three ships set out into the Atlantic. They sailed to the Strait of Magellan, led by the Pelican, which Drake now renamed the Golden Hind. Lined with glaciers, the Strait was no simple navigation problem. Narrow channels, fierce headwinds, snow, sleet and rain tore at the rigging. The Pacific Ocean came into view at the other end to much rejoicing.

The joy turned sour when they left the Strait and found themselves in the midst of a powerful storm which drove them south. The mariners struggled on as giant waves roared around and over them. They were swept far to the southeast, to the ends of Terra Del Fuego. In this terrible storm, the Marigold was lost with all 29 crewmen. The Elizabeth and the Golden Hind were separated. Drake searched but found no trace of his sister ship. He left messages in hopes she would follow him. The Elizabeth picked her way back through the Strait and returned to England under Captain Wynter. Like Magellan before him, Drake had now lost his small ship to storms and his large ship to desertion in the area of the

Strait.

Drake entered the Pacific with the Golden Hind as his only ship. The tough little ship had weathered the beating well. Drake headed north. Soon, he would be ready to tackle the Spanish galleons and shipping off the coast of Peru. The Spanish coast felt safe from attack, and the vast treasures of silver regularly crossed the Pacific Ocean. The Indians had been crushed into submission, Panama was well guarded, and the Strait of Magellan was considered too dangerous to pass. Drake had surprise on his side, but his men were hungry and exhausted. The ship, slowed by encrusted barnacles, was in need of caulking and repair.

Upon reaching the island of Mocha on November 25, Drake was again taken in by a show of friendliness, as he had been at Mogador. The natives lured a boat ashore with promises of water and food. As soon as the boat reached shore, two men jumped out to secure the lines. Thinking them to be the hated Spanish, the Indians grabbed them and showered the boat with arrows. Every man on the boat was struck several times, including Drake, who nearly lost an eye. The wounded men desperately fended off the Indians with oars and rowed for their lives. One of the boatmen died and the two who were kidnapped were never seen again. Drake immediately set out to the north.

On December 3, they anchored in another bay where a lone Indian sat fishing in a canoe. The Indian, thinking these white men to be Spanish, conversed freely with them about the port of Valparaiso where Spanish galleons anchored. Drake immediately set sail for Valparaiso, where he found the Spanish ship known as Capitana. Completely surprised, the skeleton crew gave up the ship and the villagers fled at the thought of a pirate attack.

Drake moved out to sea with this prize and some minor valuables from the village. The Capitana had a cargo of 1770 jars of wine, maize in sacks, cedar wood, and some gold. The two ships anchored on the South American coast on December 18, and a landing party proceeded to gather the provisions so badly needed. Suddenly, a large force of Spaniards and Indians approached on horseback. One Englishman was killed in the attack, but the others escaped in the boat. The horrified English watched as the body was mindlessly mutilated, beheaded, and shot full of arrows. The crew was now united in their hatred of the Spanish. This incident served warning that the element of surprise was now gone. Soon, Spanish garrisons up and down the coast would be on the alert.

Drake found another bay and set out guards. The other pinnace was assembled. As soon as the pinnace was ready, the Golden Hind was careened. Her bottom was cleaned, caulked, and repaired. The Capitana served beautifully as a rest area and potential getaway ship.

On January 19, the three vessels headed north along the coast, pillaging natives for food. Drake's destination was Arica, the silver capital of Peru. In the harbor of Arica were two ships and a well-armed welcoming committee. The silver had been hidden. Angry at this situation, Drake set fire to one of the ships and took the other. In Chile, he met with a similar disappointment and took another vessel.

Finally, he went to sea again. Now, he had to jettison his captured ships and cargo. Thus far, the big strike of silver and treasure had eluded him. His take of ships at this point would only slow his search for big treasure, so the captured ships were abandoned far out to sea. Entering Callao, his luck proved no better. By now, two warships with 120 soldiers were in pursuit of the English buccaneers. With this fresh information, Drake was soon chasing another Spanish treasure ship, the Nuestra Senora, nicknamed the Cacafuego (Spitfire).

After capturing a small Spanish ship, Drake had been able to learn about Spanish shipping in the area by interrogating its pilot. He had terrorized one small vessel after another, before he spotted the large, fully armed Cacafuego. Hiding the pinnace behind the Golden Hind, he dropped water filled wine jars on a line astern to create drag. Then he set full sail. The Golden Hind now appeared to be a fully laden and slow moving merchantman. Meanwhile, the cannons were rolled out, loaded and primed. Men were armored and armed. The ship bristled, ready for action.

When she came alongside, the Cacafuego hailed the Golden Hind.

"Strike sail, Mr. Juan de Anton," Drake greeted the captain, "or we will send you to the bottom."

"What old tub is ordering me to surrender?" the captain replied.

"Come aboard yourselves and strike sail," he challenged.

A cannon blast tore off the Cacafuego's mizzen mast and a fusilade of arrows and shot roared across her decks. From the other side a band of armed brigands boarded out of the pinnace. The captain and crew of the Cacafuego were quickly made prisoners and courteously treated. Inspection proved that the ship was the big haul Drake had been seeking. The ship was emptied of 1300 bars of silver, 14 chests of coins, gold, jewels, and other vast riches and treasures. By the time the Cacafuego was released, the expedition had become a highly profitable one.

The insatiable Drake wanted more. He sought revenge against the Viceroy of New Spain and Mexico. This was only a symbol of his anger against the Spanish. He captured a bark which he exchanged for the pinnace, but his raids in Mexico were less than satisfying, and not monetarily profitable.

By the time April came, the heavily laden Golden Hind had left Gualtiuco on the Mexican coast with food and water for a fifty day voyage. To try to and take the Spanish route from Acapulco to Manilla would place them in that area during typhoon season. To head south to the Strait of Magellan would inevitably bring them into contact with the Spanish pursuers, and by now the Strait was blocked and guarded. To wait three months in enemy waters was equally dangerous. Drake had no choice but to head north and try to discover the northern route east and home.

Drake headed northwest. Maps of the day indicated that the coastline headed this way. For two months, they sailed through storms and fog. Provisions ran low, and still there was no sight of land. Drake turned east. The coast, which he finally reached on June 5, was shrouded in bad weather. From their first landfall, they moved southward until they found safe anchorage at San Francisco Bay, which Drake claimed for the Queen. Here the Golden Hind had to be careened and cleaned. The men built stone structures and befriended the natives. This provided an opportunity to rest and reprovision for the long inevitable journey across the Pacific. The treasures were carefully stowed, and the Spanish bark was abandoned. They spent a month of hard work in this far away hidden bay, safe from Spanish searches, unfriendly Indians, storms, and the rigors of sea duty. Now it remained for the men to head home and cash in on their wealth. Since no northern passage was available, they were committed to circle the globe to survive.

The Pacific had lost much of its terror since Magellan had crossed it. Spanish ships regularly traveled routes such as the westward route from

Mexico to Manila. On July 25, Drake headed into this fearsome stretch of sea. His ship would be slowed by over forty tons of treasure, a full load of food and the equipment for the voyage. She was sturdy and seaworthy, and normally a fast vessel. The proper time of year was at hand.

Drake had an advantage over Magellan. He knew he would be at sea for three months, and he provisioned accordingly. He steered southwest, and warm trade winds carried them over the calm sea. Unlike Magellan, there were no other ships in the fleet to break the monotony of the empty seascape, and they went without a break in the empty horizon for 66 days.

On September 30, they spotted land, probably the Carolines. They were surrounded by Polynesians in polished dugout canoes. Since the natives were more interested in accepting than exchanging food, they had to be dispersed by discharging a demiculverin. Drake carefully picked his way through the islands to Mindanao in the Philippines. They had to be wary. They were Englishmen in a world under Spanish and Portuguese influence. There were other pirates. Arabs, Chinese, natives, and Moors had to be avoided.

While moving south in search of the Spice Islands, and the route home, Drake encountered a Portuguese ship. He needed no more booty and did not want a fight. The Portuguese, thinking the Golden Hind to be Spanish, pursued her. Finally, Drake made the ship battle ready and ordered the Portuguese ship to strike sail. The Portuguese had forts and garrisons in the area, and this ship was well armed and ready to take on pirates of any nationality. The Portuguese would not bluff, so Drake, realizing how much he had to lose, turned and ran. To everyone's relief, they were not pursued.

The natives, on the other hand, turned out to be allies. Upon reaching Ternate, they found a sultan named Baab. The Portuguese had been overthrown here in 1575, and the Englishmen represented hope of a power which could aid against Spain and Portugal. Drake and the sultan received each other graciously, with lavish displays of gun salutes and music. In this rich trading center, the sultan lived amid trappings of great wealth from all over the world. Drake took on six tons of cloves and established relations, for the first time, between England and the Spice Islands of the east.

Getting home was now all important. Drake left, and when offered an opportunity to go to China, turned it down. He had barely sixty men left. All were tired and many were sick. He searched out an island where the ship could be careened one more time for the homeward voyage.

On December 12, the ship continued on, picking her way past the coral reefs. In the dark of night on January 9, she struck one, heeling over dangerously. The wind that had driven her broadside onto the reef now held her there. Sensing that they were in as great a danger as they had yet seen, the listing deck filled with frightened, scrambling sailors. Drake reacted quickly to calm the men. He called the crew together for prayer. Having first appealed to God, Drake began action to free his ship. He knew they were in danger from any onslaught of wind or wave, and that if she were not freed they might be forever at the mercy of local natives.

The ship's boat went out to take soundings, searching for some rock on which to secure the anchor and pull the Golden Hind off of her precarious perch. On one side, they found bottomless ocean, and on the other, only seven feet of water, a disappointing level against the ship's draught of 13 feet.

Drake ordered the ship lightened. Cannons, hatches, precious cloves and food were thrown overboard. The ship held fast, plastered by the heavy wind. Drake considered. There was no leak or damage and the gold and silver treasure was tucked safely in the bottom as ballast. He called the men together. The buccaneer looter turned Puritan Englishman again, calling for them to commend themselves to God's mercy. Francis Fletcher preached a sermon, evoking the need for forgiveness of sins at this critical hour, especially for the killing of Doughty 19 months earlier in San Julian. Drake and his men listened in conscience stricken silence.

In midafternoon, the wind dropped. The ship heeled over in the other direction and was suddenly free in deep water. Scared saints returned to pirating sinners, and Drake ceremoniously excommunicated Fletcher, banning him to the area before the mast on pain of hanging. This placed him in the front of the ship away from the officers and gentlemen for the rest of the voyage.

The Golden Hind persisted onward, threading her way through the islands and reefs. They found an island where the natives were friendly and they were able to stock up on meat, lemons, and coconuts. More anxious than ever to deliver his precious cargo, Drake did not tarry long. He sailed for another month before stopping on the southern coast of Java to prepare for the long journey across the Indian Ocean.

At Java, Drake befriended a rajah. The rajah was sumptuously entertained aboard the Golden Hind and the decked out English gentlemen were entertained lavishly on shore. Once again, the work of cleaning and reprovisioning began. The coming journey would be as arduous as crossing the Pacific, for like Del Cano, Drake would have to avoid the Portuguese and head for the Cape of Good Hope. The men filled water casks, salted and packed meat, and repaired the ship. Interrupting the calm which they were enjoying, Drake simultaneously discovered a Portuguese spy among the natives and received word that three vessels were approaching along the coast.

Quickly, the Golden Hind took her leave, sailing into the Indian Ocean on March 26. With good trade winds and fair weather, she was able to make the crossing in 56 days, compared to Del Cano's 136. The weather held, but it took over three weeks to round the Cape. Thirst became the severest problem. Only a rainstorm allowed them to have enough drinking water to reach Sierra Leone. A meager eight pints of water remained when they arrived on July 22.

Home was a relatively short distance now. Their cycle of filling water casks and foraging for food was repeated one more time. The men were treated to a view of elephants for the first time in this African land. In two days, they set out again.

On Monday, September 26, 1580, they returned to their homeland, and the reality of politics. Spain was demanding the return of the ill-gotten gains, registered on the amount of £332,000. In strictest secrecy, the Golden Hind was unloaded. Most of Drake's treasure was unregistered. The tonnage of wealth enriched the Queen's treasury, returned to her backers £47 for every pound invested, added to the wealth and fame of Francis Drake, compensated the surviving crewmen, and reimbursed Spain for the full amount that had been requested.

Both Magellan and Drake lost four out of five ships. In Magellan's voyage 18 out of 277 men returned with the Victoria. In Drake's voyage, 58 out of 164 men returned. Magellan's expedition had lost 94% of its complement, roughly 61% to death and 34% to desertion. A small percentage had been arrested. Drake lost only 64%, roughly 34% to death and 30% to desertion. Early circumnavigations were monetary successes even with the considerable loss of human life, so were considered to be worthwhile by their respective governments.

This second circumnavigation of the globe between December 12, 1577, and September 2, 1580 took two years, nine months, and 13 days, some two months and four days less than Del Cano's initial record trip, which had stood for 58 years. The next circumnavigation would come along much sooner, but the record would not pass from the hands of the English just yet.

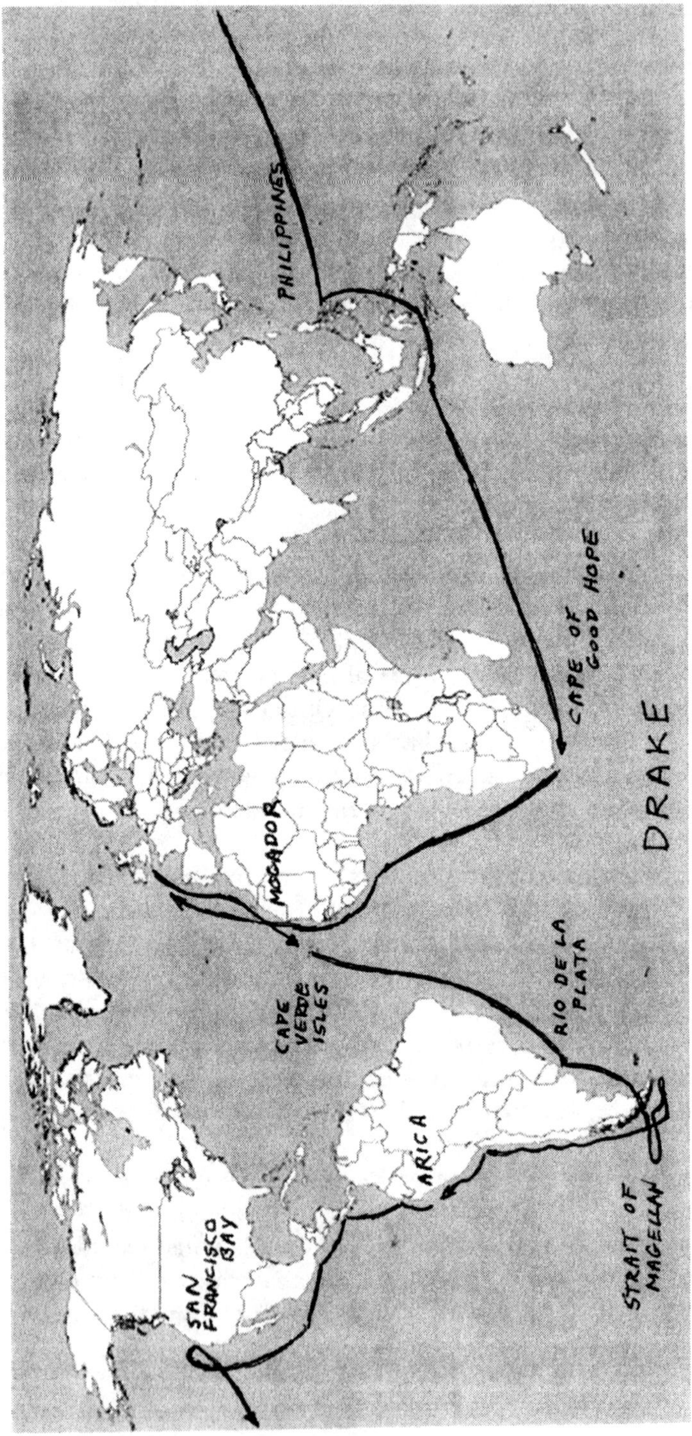
DRAKE
PHILIPPINES
CAPE OF GOOD HOPE
MOGADOR
CAPE VERDE ISLES
RIO DE LA PLATA
STRAIT OF MAGELLAN
ARICA
SAN FRANCISCO BAY

CHAPTER 3 - CAVENDISH

Relations between England and Spain did not improve in the years following Drake's return. Drake had become an extremely popular folk hero. When there was a lessoning of tensions, he would slip into obscurity, but when diplomacy between the two powers broke down; his example would shine forth as representative of English resourcefulness and grit. England called on Drake whenever there was a need for a first rate seaman. He led successful raids in the West Indies, and when Spain assembled an armada in 1587, Drake boldly sailed a squadron into Cadiz harbor and burned 32 Spanish ships at anchor.

While much has been written on Magellan and Drake, the record breaking circumnavigators that came after them became increasingly obscure. Most of our knowledge of Thomas Cavendish and his voyage is assembled in *Hakluyt's Voyages*, with the best account being that of Francis Petty, who accompanied the expedition. Unlike Drake and Magellan, Cavendish intended from the start to circle the globe, and provisioned his fleet for at least a two year journey.

Cavendish's motivations are not clear, but it is known that he had invested in Sir Walter Raleigh's unsuccessful attempt to colonize Virginia at Roanoke in the New World, so there is a chance that the motivation was more financial than patriotic. Cavendish had been born to gentility and came to his position not by intense effort, but by the rigid social structure of the day. With a little help from his friends, he secured the backing of the court and the Queen for his proposed venture. He assembled a fleet in the remarkably short period of six months.

The fleet for this privateering expedition against Spain was the smallest yet, consisting of the 120 ton Desire, the 60 ton Content, and the 40 ton bark, High Gallant. The total fleet was 220 tons compared to Drake's 295 tons and Magellan's 480 tons.

Loaded with all manner of furniture and food, the fleet weighed anchor at Plymouth on Thursday, July 21, 1586, and sailed south.

A few days later, they came across five Spanish fishing vessels. For three hours, they exchanged shots, but broke off their attack at nightfall.

On August 26, after a month at sea, they anchored at Sierra Leone, where Drake had restocked his water supply on his way home. The Content led the way, taking soundings as they entered the harbor. The town was a well-kept place with mud buildings, clean yards, neat streets and over a hundred houses. At first, the men enjoyed friendly relations with the Negro populace, and learned of a Portuguese ship further up the harbor. The Hugh Gallant searched a little farther, but found the soundings unsatisfactory and returned. Then the men found a Portuguese citizen hiding in the bushes. He repeated what they had already learned, that it was too dangerous to take a ship farther into the harbor.

This seemingly innocuous event changed their relationship with the locals, and Cavendish sent a party of 70 armed men into the village to burn and pillage. They destroyed three houses and found very little plunder. The Negros fled, shooting arrows at them from a distance. Later, as the men were washing their shirts, they were rushed by the locals. One soldier, named William Pickman, was struck in the thigh. He broke off the arrow, leaving the arrowhead in his flesh. Fearing the ship's doctors, he said that he had pulled it all out. The next day, his belly and groin were swollen and black, and he died shortly thereafter. The expedition had suffered its first fatality.

On September 6, the three ships left Sierra Leone and crossed the Atlantic. No longer did Englishmen fear the great South Atlantic, nor the Great South Sea, as the Pacific was known. By the last of October, they had reached Brazil. They landed on the island of San Sebastian and set up a forge. The Coopers made hoops and casks, and the men assembled a pinnace which they had brought along.

On November 23, they left San Sebastian headed south along the

Brazilian coast. On December 17, they entered a harbor which they called Port Desire. Drake had visited Port Desire on his voyage, and had named it Seal Bay. Today it is known as Puerto Deseado, Argentina. The area abounded in seals, which yielded meat, valuable skins, and blubber. The harbor was an excellent shelter, and allowed them to clean and repair their ships. The natives were unfriendly, however, and attacked, wounding two members of the crew. The soldiers who followed this raiding party managed to measure one of the footprints, finding it to be 18 inches long, and giving history an accurate measurement of Pigafetta's giants.

Leaving Port Desire, they sailed to the Strait of Magellan, entering it in early January. They had timed their passage well, for it was now midsummer in the southern latitudes. Once in the Strait, Cavendish found the tragic survivors of a Spanish garrison which had been placed there three years earlier to guard against foreign intrusion. In 1584, four hundred Spaniards had established a fort at the best location in the forbidding straits. They had food, water, and mussels. They erected a city with forts, complete with a cannon, a church, and a gibbet for executions. There was no good cropland, however, and the long cold winters had taken their toll. When Cavendish found them, most had died of starvation. Only 21 men and two women remained alive out of the original four hundred. Cavendish offered to take them to Peru. The Spaniards withdrew to consider the offer, but Cavendish had to leave before the skeptical Spanish were able to make their decision and return to the English. Only two of the Spaniards survived, one who was already on board when Cavendish left, and one picked up by another English ship in 1589.

A crew member, Francis Petty, faithfully recorded all of the deaths of the expedition members as the expedition progressed. One of them, a carpenter of the Hugh Gallant, died and was buried in a bay they called Elizabeth Bay on January 21. On February 24, they exited the Strait into the Pacific.

On the first of March, a storm took the ships north. The Hugh Gallant was separated from the fleet. For three days, the exhausted men battled the raging sea in the leaking bark, afraid to sleep. After the storm abated the Hugh Gallant sailed for fifteen days before finding the other ships close to the island of St. Mary, near Concepcion, Chile. The Indians, thinking them to be Spanish, offered gifts of corn, wheat, and barley. They were also able to reprovision here with salted pork, dried fish, maize, and tubers.

Cavendish was now ready to begin his buccaneering raids against the Spanish seacoast, but they were not the easy prey that Drake had met. Armed and ready, they proved formidable enemies.

On April 1, 1587, several men went ashore to fill water casks. While they worked, 200 Spanish horsemen descended on them. Twelve men were killed or captured in the skirmish, and English soldiers fought for an hour before driving off the Spaniards. Petty dutifully listed the casualties noting that six were from the Desire, two were from the Content, and four were from the Hugh Gallant. To posterity, these facts mean little, but to the families waiting at the end of the voyage, it would be some comfort to know where their loved ones were lost.

Leaving the area, they came upon a small bark, which they captured and named the George. In the harbor of Arica, the pinnace took a ship of nearly 100 tons. When the Spaniards refused to ransom the ship, which carried no cargo, the English burned it. Cavendish was not making much of a name as a pirate, but he continued on, heightening his efforts.

The fleet separated, but this was of little help as the towns were prepared with garrisons waiting and their treasures hidden. The English did finally manage to take some Spanish ships. One was 300 tons, but the captured cargos were only corn meal, sugar, hides, molasses, and timber.

Upon reaching the island of Puna, Cavendish managed to sink a 250 ton ship. The fleet rendezvoused, and Cavendish ordered the Desire careened and scraped. As time passed, the English allowed their guard to become lax. During the night of June 2, the Spanish and Indians slipped onto the island. Of the 20 men caught on shore, 12 were killed in the ensuing battle. That same day, Cavendish sent 70 reinforcements to fight the Spaniards. In retribution, they set fire to the town of some 300 houses, burning it to the ground, along with four ships being built there. The expedition had now lost 26 men. After leaving Puna, Cavendish had the Hugh Gallant scuttled for lack of sufficient crew.

The fleet continued up the coast. They burned and sacked the port of Guatulco, in Mexico, on June 27. They looted their way up the Mexican coast, still finding no gold or silver. Cavendish abandoned his captured bark, George, and headed the Desire and Content for Baja California, to wait for a Spanish galleon from the Philippines. His cruise up the western coast of South America and Mexico had been discouraging, and when they spotted a sail off Cape San Lucas on November 4, they were eager to give chase. The ship they were chasing was the 700 ton ship, Santa Anna, much larger than the Desire.

The Santa Anna had been spotted about eight in the morning. In the afternoon, the Desire drew alongside the big ship and fired a broadside of heavy ordinance and a volley of smaller shot. They approached for boarding only to be met with a shower of lances, javelins, rapiers and stones. Two of the Desire's crew were killed in the barrage and four or five were injured. Cavendish backed off and let fly another broadside at the Santa Anna. Much of the heavy shot struck beneath the water line. A third broadside tore through the big ship, maiming and killing many of her crew. With the Santa Anna now in danger of sinking, the captain put out a flag of truce.

There were 190 men and women aboard the prize. She carried a cargo of 122,000 pesos of gold, silks and satins, fine food and wines. Cavendish graciously entertained the Spanish officers and took great care to provide for the prisoners. They were set ashore with food, sails to make tents, enough planking to build a bark, and sufficient weaponry to defend themselves against the Indians. There was a nearby river for fresh water and fish. The Englishmen set about transferring the treasures to their ships. On November 17, they celebrated the anniversary of the coronation of Queen Elizabeth. They put on an impressive display of fireworks for the benefit of the Spanish, discharging all of the ordinance on both ships.

For his journey home, Cavendish took with him a Spanish pilot, a Portuguese pilot, two Japanese, and three young Philippinos, all of which should be useful in the homeward navigation. Feeling he had made the big haul, Cavendish was now ready to return to England without further delay. They set fire to the Santa Anna and sailed. The Santa Anna drifted ashore where the Spaniards were able to extinguish the flames and make repairs enabling them to sail her to Acapulco with their tale of woe. Unlike Drake, who could successfully burn an entire Spanish fleet in Cadiz, Cavendish didn't even succeed in burning one vessel.

Cavendish had already lost 28 men of the original 123, but another tragic loss lay ahead as the two ships sailed into the vast Pacific. The Desire left the Content astern in a fair wind, thinking they would quickly catch up. Nothing was ever seen of the Content or her crew again. The Desire was alone to finish her voyage around the world and to carry the booty. The Pacific crossing was only 45 days before a fair wind, and they reached the Ladrones on January 3, 1588. The islanders gathered around them in canoes and crowded so close that the canoes overturned or broke against the sides of the Desire. The travelers traded iron for food by putting a line over the side to the canoes. The natives detached the iron and fastened bundles of roots or plantains. Finally, Cavendish was forced to order the arquebuses fired to make the canoes leave so the Desire could continue.

The Desire sailed on to the Philippines, which were infested with Spanish. Manila was to be avoided for there were over six or seven hundred Spanish to contend with. From here, each of the early circumnavigators had threaded his way south through the many islands and coral reefs to the open seas. Cavendish worked the Desire south, stopping at an island off Luzon and again at Capul. He concluded treaties with native chieftains, assuring them that he was not Spanish.

The heat was oppressive and some of the Europeans fell victim to its choking hold. On February 14, a cooper died of the sickness and on the 21st an officer died of the fever. Petty was among those who became deathly ill from the tropical heat. Cavendish bypassed the Spice Islands, which had greatly enriched Drake's expedition. Like Drake, he stopped at Java, where the Desire was filled with provisions for the long trek across the Indian Ocean to the Cape of Good Hope. On March 14, the eager and homesick mariners set out toward the southern tip of Africa. The rest of March and all of April were spent traversing this sea and on May 11, Africa came into view.

Rounding the Cape, Cavendish sailed to St. Helena. This island sat by itself in a strategic location in the middle of the South Atlantic. The Desire took on wood and water and more provisions.

The world had become increasingly civilized as the years had gone by. Missing from the narratives of this voyage are the tales of scurvy and starvation. The Desire reached home without having suffered a serious mutiny. Perhaps this was because the nature and objectives of the voyage were known from the start, and Cavendish had the advantage of a mostly English crew, and the rank of birth. He did not have the ability of Magellan and Drake as a seaman, however, and he returned home with about half of his original complement. His voyage was not as profitable as the two before him. It was sufficient only to cover the costs of the expedition.

Cavendish had completed the third circumnavigation of the world in the record time of two years, two months, and ten days. Drake's record had stood for a scant eight years. The English would retain the title for 29 more years before a new nation of people entered the circumnavigation scene.

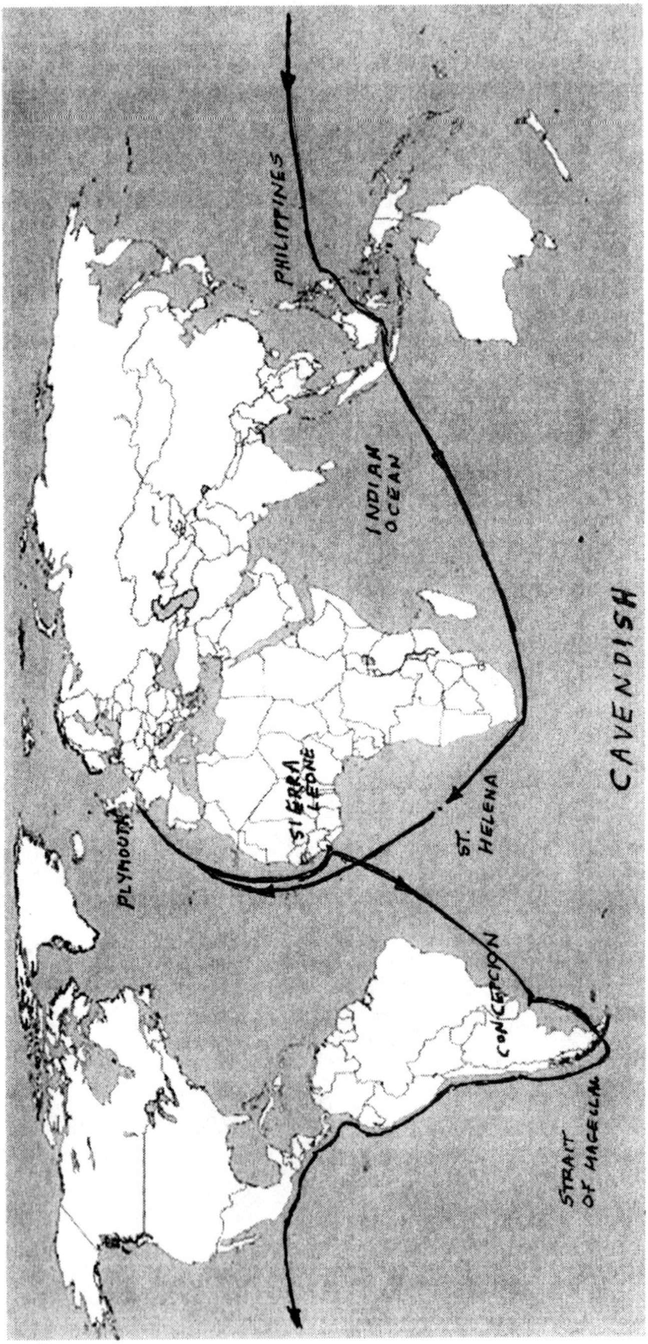
PHILIPPINES
INDIAN OCEAN
CAVENDISH
ST. HELENA
SIERRA LEONE
PLYMOUTH
CONCEPCION
STRAIT OF MAGELLAN

CHAPTER 4 - SHOUTEN-LEMAIRE

Cavendish died in 1592, in a disastrous attempt to repeat his earlier voyage. Drake also died while on a voyage in Central America in 1596. Sailing the earth's vast oceans was by no means safe or simple. Voyage after voyage ended in failure. Scurvy and hunger, storms and calm, disease and hardship still plagued the sixteenth century seaman, but by the end of the century far more was known of our planet than had been known at the beginning.

By the latter part of the century, a new force had rapidly developed in the East Indies. The low countries of Europe, Flanders and Holland, had shaken free from Spanish domination. Cities like Antwerp, Amsterdam, and Hoorn had developed maritime abilities and spread their sails far beyond fishing and hauling wool. Above all, the Dutch were astute merchants, determined to cash in on their share of eastern riches.

Competition between nations and even companies depended on force. Ships with cannons and soldiers carried on trade. The Portuguese had gone east past the Cape of Good Hope, and by-passed the Moslem and Arab traders. The Spanish had gone west around South America and by-passed the Portuguese. The English were making their way as pirates raiding the other two nations' ships. No longer would a ruling Pope divide the spoils in an increasingly Protestant world. Brutal exploitation of newly discovered lands led to resentment and opened doors to more proper trade and new alliances.

Moving carefully, the Dutch sent expeditions east with varying degrees of success. The cutthroat competition between a proliferating number of

Dutch companies in the East Indies began to make trading difficult for all. Finally, at The Hague in 1601, Consolidated National Company was created, governed by 17 directors from Amsterdam, Zeeland, and other Provinces. A 321 year monopoly charter was granted to this newly formed United Dutch East India Company (Vereenigde Oost-Indische Compangnie), known as VOC. So successful were the Dutch that these areas of trade would become known as the Dutch East Indies for centuries.

The Dutch had completed one slow circumnavigation of the globe, and the VOC had another in process when a rival expedition was put together. Although the expedition itself was destined to not complete its circumnavigation, when some of its members returned to Europe they would have set a record that was to stand for over a century.

Isaac LeMaire of Amsterdam was determined to find a new route to the Indies. Although the VOC now controlled the Strait of Magellan and the Cape of Good Hope routes, anyone who could find a new route would have exclusive rights to the first four voyages using any new route or for any newly found lands. Willem Schouten and Isaac LeMaire believed that a strait lay south of the Strait of Magellan. The expedition, organized in 1614 and 1615, was to seek that second passage through the New World, and perhaps find that large undiscovered southern continent, Terra Australis Incognito.

LeMaire outfitted two ships, the 220 ton Eendracht with a crew of 65, and the 110 ton Hoorn with a crew of 22. Isaac LeMaire's grown son, Jacob, was placed as supercargo, the owner's representative in charge of cargo. Willem Schouten was captain of the Eendracht and his brother Jan captained the Hoorn.

The two ships hoisted anchor at Texel on June 14, 1615, with only the three top officers knowing their proposed route. The other mariners had signed on with the understanding that they would follow the route chosen by their leaders.

Three days later, they stopped at Dover to hire an English gunner, then on to Plymouth to hire a carpenter. Each crewman was rationed a tankard of beer a day, four pounds of bread, and a half pound of butter a week. The plan was to obtain fresh fruits and vegetables as they went. They reached Cape Verde Islands on July 23, but were unable to obtain the desired supply of lemons. By late August, after spending much time becalmed off the coast of Africa, they put in at Sierra Leone, with several of the men suffering from scurvy. Finally, they reached a section of the coast where they found a village of Negroes with an abundant supply of lemons. In a place with plentiful lemons and scarce iron goods they purchased some 25,000 lemons and a quantity of fish.

Contrary winds made progress for the two ships extremely slow, and they were unable to put much distance between themselves and Africa after sailing. On October 5, the Eendracht suddenly shuddered and a great noise was heard forward. Rushing forward, Shouten could only see a vast quantity of blood in the sea. Only later upon beaching their ship, were the men to learn that they had been struck by large swordfish or narwhale. The large horn described as the thickness of an elephant tusk was embedded through two stout fir planks and an oak plank, almost penetrating the ship's side.

Upon crossing the equator on October 28, Schouten and LeMaire announced their route to the crew. The crew seemed pleased. On December 6, the ships spotted the coast of South America and headed for Port Desire.

They anchored in some 20 fathoms, but the anchors did not hold and the ships drifted shoreward. The vessels lodged precariously on the rocks when the tide went out. They managed to refloat both ships when the tide returned, but only through the good fortune of fair weather. The men found penguins and penguin eggs, but little good water. They did however; uncover the graves of people apparently ten or eleven feet tall. Once again, we see recorded evidence of a race of giants in South America.

The ships were beached, deliberately this time, and careened. The hulls were cleansed of barnacles by fire. While the Hoorn was beached, the tide went out, leaving her 50 feet from the water. Suddenly flames leapt into the rigging. Before men could do anything, the conflagration was out of hand, and they could only watch her burn. The gunpowder exploded, scattering timbers, cargo, and tools over a wide area. All night her crew watched her burn. Finally, the charred remains were broken up with her anchors, guns, and salvageable timbers being loaded onto the Eendracht.

Before leaving Port Desire, the crews discovered an ample supply of fresh water. In one day, ten tons of water were brought aboard the Eendracht. On January 13, 1616, seven months after leaving Texel, Schouten and LeMaire headed south into the cold, tempestuous seas. Guns and everything that could be taken from the decks were securely lashed below. The land they passed was mountainous and covered with snow. Land, perhaps the great Southern Continent, blocked their way to the south. They continued along the coast of Tierra Del Fuego south of the Strait of Magellan, until they came to a channel between two land masses. The one to the west they knew. The one to the east they called Staten Land in honor of the states of the Netherlands. When they emerged from the channel, later named the LeMaire Strait, they could feel the big waves and knew they could reach the Pacific. This mountainous, snow covered cape they named the Cape of Good Hoorn in honor of the same city their lost ship had been named after. These were not the first men in the area. It was here that Drake's ships had been blown and separated in a great storm that greeted his entrance to the Pacific. From now on, this fast, open, and much less hazardous route would be used, until the Panama Canal was cut centuries later.

These were cold and debilitating waters and the Eendracht headed north as quickly as possible. It was not fast enough to save Jan Schouten, Willem's brother, who died on April 9, after a month of suffering and sickness. He was the expedition's second casualty. The Eendracht sailed rapidly in the Pacific, stopping at an occasional island. They experienced the usual exuberance of the natives, who would try and tear nails from the timbers of the ship. At each stop, they filled water casks and reprovisioned with whatever each island had to offer. On one such island, the men returned covered with flies. So black with insects were they, that their faces and hands could not be seen.

The flies stayed on the ship, covering everything. The mariners swatted and killed, but they were in all they ate or drank. For days, the pesky insects infested the ship until a fresh wind came up, and the flies took leave with the wind. The crew called the island, "Island of the Flies".

On May 8, they sighted a sail, rapidly closing from the south. Schouten fired a shot across her bows, the international signal to strike sail and be identified. But this was a native ship, unfamiliar with such a signal, and she attempted a desperate escape. The Eendracht launched her boat with ten musketeers, who fired several times at the fleeing ship. As they boarded her, several natives dove overboard in terror, some drowning.

The Dutchmen did their best to undo the damage, and plied the remaining frightened men, women, and babies with gifts before sending them on their way wailing with grief over their lost loved ones. LeMaire was astonished to watch these people drink salt water directly from the sea.

The Eendracht continued her westward journey, still searching for Terra Australis. The Dutchmen found an island where the natives seemed friendly and willing to trade. Nearly a thousand of them gathered, and the crew bartered and prepared to sail. The native canoes then gathered into battle formation, and attacked the big Dutch ship, pelting them with stones. A few rounds of nails from the big guns allowed the Eendracht to escape.

The expedition did not have good luck with natives until they reached islands which they named Hoorn Islands, where a friendly but fearful king provided them with abundant food and feasts. Onward the ship sailed, first fighting off attacking canoes,, then passing an active volcano belching fire and smoke. As they wandered through these islands, they encountered an earthquake, thunder and lightning and a rain so heavy they had never seen anything like it. All of these phenomenon could have been manifestations of the eruption of a volcano spewing its pollutants into the air, but there was probably not sufficient knowledge of weather for the Dutchmen to know this.

The Eendracht picked her way through the difficult areas of the Spice Islands searching for the Dutch controlled port of Ternate. On September 17, she anchored safely at Ternate in one of the fastest and safest outward trips ever. Here, 15 men transferred their employment to the VOC. This circumnavigation had now crossed paths with another Dutch circumnavigation, by Joris van Spilbergen. Spilbergen, working for the VOC, had left Texel on August 8, 1614, with six ships. One of these ships, the Morning Star, met the Eendracht near Ternate in the Spice Islands. Having left nearly ten months before the Schouten-LeMaire expedition, Spilbergen had passed through the Strait of Magellan, and had been in the Dutch East Indies for several months. The Morning Star accompanied the Eendracht part way to Djakarta in Java, where two more ships of Spilbergen's fleet were taking on cargo for the return trip.

Schouten and LeMaire proudly related their stories of discovery to the local representative of the VOC. He had little interest in hearing of a competitor's success, nor did he believe the claims. He ordered the Eendracht and her cargo impounded. Thus, the VOC took possession of the good, sturdy ship that had taken Schouten and LeMaire halfway around the world. The crews, deprived of a way home, mostly enlisted in the service of the VOC. Schouten and LeMaire were placed on board Spilbergen's two ships, the Great Sun and the Half Moon, which sailed for home by way of the Cape of Good Hope on December 14, 1616.

On Java, the Schouten-LeMaire expedition had its third casualty when a sailor died, but it was not the final loss. Isaac LeMaire became ill shortly after the humiliating loss of his ship at the moment of triumph. He died at the age of 31, a little over a week from Djakarta, and was buried at sea. Spilbergen's two ships were separated in January. The Great Sun stopped at Mauretius to fill her casks with water, and then passed around the Cape of Good Hope though records indicate that they didn't see it. On March 30, the Great Sun reached the island of St. Helena, where she met the Half Moon once again. Together, the heavily laden vessels sailed on to Texel, where both expeditions ended together.

The expedition of Schouten and LeMaire was the first record breaking circumnavigation to be a financial failure. The aging Isaac LeMaire had lost his ship, and his son. Although the courts later ordered compensation for the confiscated ship and cargo, the VOC ignored the court orders and Isaac could not enforce it. The men of the Schouten-LeMaire expedition had circled the globe in two years and 17 days, and returned in someone else's ship. It would be over a century before anyone equaled their record. Willem Schouten was able to publish his story, and the world had a new route around its circumference. Never again would ships have to struggle through the dreaded and hazardous Strait of Magellan and the new path would bear the name of its discoverer, the LeMaire Strait. Cape Hoorn, later Cape Horn, would also become a legend.

Even the breaking of Schouten's record a century later would not wrest the record from the Dutch. The search for trade routes and piracy had motivated the first four record breaking trips. Many trips around the world would follow in the seventeenth century, but the next record would be set in the eighteenth century, in a rather unsuccessful voyage of discovery.

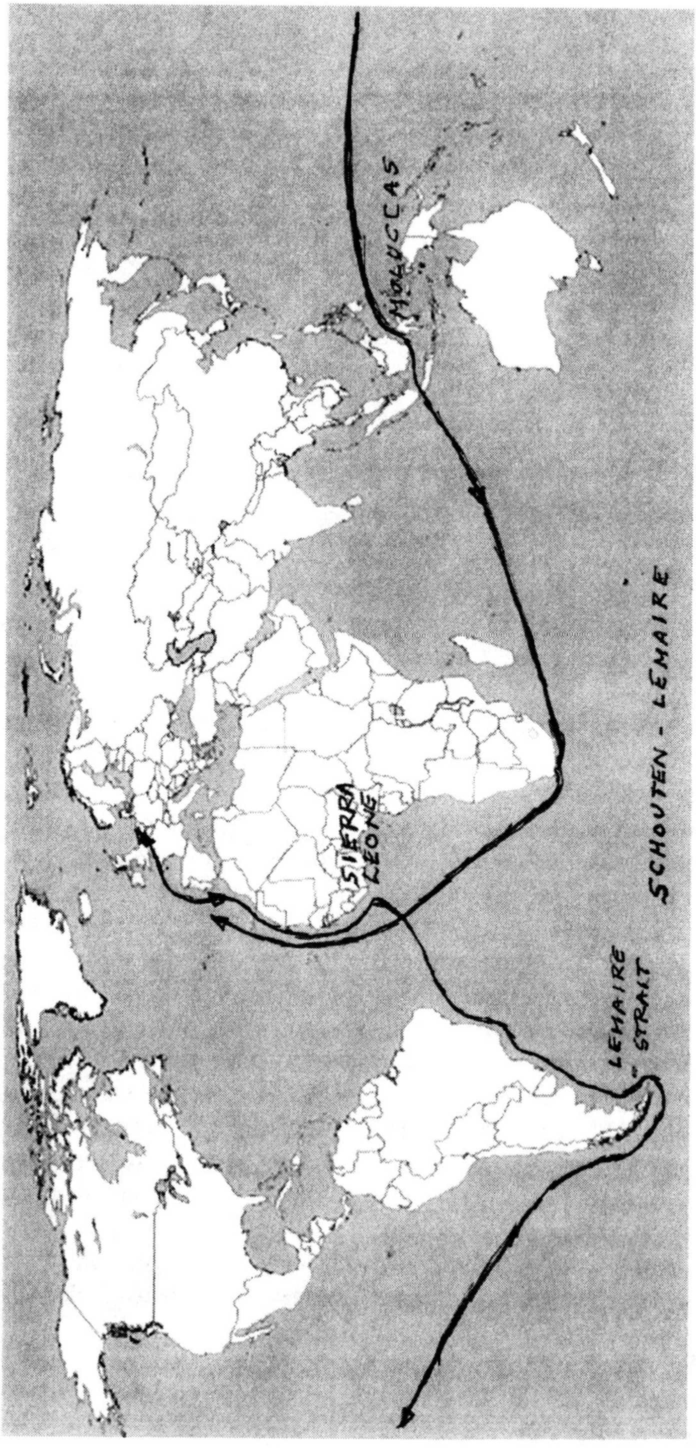
MOLUCCAS
SIERRA LEONE
SCHOUTEN - LEMAIRE
LEMAIRE STRAIT

CHAPTER 5 - ROGGEWEIN

As was becoming an international sport, the Dutch West India Company had been formed in 1621, to carry on economic warfare against Spain and Portugal in the West Indies, South America, and Africa.

In 1669, Jacob Roggewein's father had asked the West India Company for three armed vessels to explore the Pacific. Touchy Spanish-Dutch relationships kept the project on the shelf and the elder Roggewein, upon his death, passed his dream on to his son. Jacob promised to execute his father's plan, but he was 62 years old before the proper opportunity came. By that time, he had made several voyages to the Indies, and had served as a judge in the Batavian Justice Court.

Many voyages around the world had taken place in the 104 years since the Schouten-LeMaire expedition, and many sailors could claim to have made the circuit in the course of their careers. These were time consuming voyages, however, undertaken for financial gain and not for speed. Although the voyage of Schouten and LeMaire had shown that such trips could be taken without heavy loss of life and ships, a sailing trip around the world was still fraught with great dangers. Storms, disease, hunger, loss of wind, and unfriendly natives plagued sailors throughout the sixteenth, seventeenth, and eighteenth centuries. They also had to contend with pirates and rival nations for their path in the sea.

The New World was rapidly being explored and colonized. The natural harbor of Rio de Janeiro, where Magellan's crew cavorted with the natives, was now a thriving Portuguese city. As the Spanish and

Portuguese carved out permanent colonies in South America the French and English were gaining solid footholds and pushing inland on the wild and rugged North American continent. The Dutch had staked their claim on the Spice Islands, while Spain settled Manila and the Philippines. India and Arabia were still solid trading centers, while Africa, except for the oceanic fringes, remained dark and unknown. China remained aloof, and awe inspiring to the Europeans, who only glimpsed this vast domain with its rich and exotic culture.

Open and unexplored were vast sections of the Pacific. One by one, explorers in tiny ships had traced thin lines of knowledge across this open sea. Jacob Roggewein was in search of new lands in the South Pacific. Undiscovered lands had been a source of wealth to the nations which charted them. These lands had yielded up spices, gold, silver, and slaves to their new claimants. If nothing else, they could grow sugar, make rum, grow tobacco, or be a source of timber.

The fleet of Jacob Roggewein assembled in 1721, and consisted of the Eagle with 36 guns and 111 men, the Tienhoven with 28 guns and 100 men captained by James Bauman, and the galley African with 14 guns and 60 men under Captain Henry Rosenthal. The ravages of scurvy and its general causes were well known by this time. Wooden boxes of earth were placed along the bulwarks of the three ships. The idea was to grow fresh vegetables along the way to combat the dreaded disease. Instead of casks filled with biscuit and bread, which could mold, turn stale and become worm infested, ovens were brought on board and flour was placed in the stores. Fresh bread was to be baked along the way.

On August 11, 1721, the ships with their complement of over 270 men left Texel. Sailing across the Atlantic, the fleet stopped at Rio de Janeiro before exploring in search of new islands in the area. They continued south, passing through the Strait of LeMaire, and were carried as far south as 62 1/2° southern latitude. The Tienhoven was separated from the other two ships on December 21. The Eagle and African moved north to the coast of Chile and cast anchor opposite the deserted island of Mocha. Quickly they moved to Juan Fernandez Island, where they found the cabin of Alexander Selkirk. Selkirk was a Scotch seaman who had been marooned for 4 1/2 years on the deserted island. During this time alone, Selkirk had survived by hunting native goats and growing vegetables previously sown by the crew of a ship. He had been rescued in 1709, by Woodes Rogers on his voyage around the world. The story of Selkirk inspired Daniel DeFoe to write Robinson Crusoe. The

Tienhoven was reunited with the fleet while they harbored here.

Before March ended, the ships left the anchorage and steered west-northwest. After several days, they sighted an island. It was Easter Sunday, April 6, 1722. Roggewein named the island Easter Island. The shore was lined with tall statues, strange enigmatic guardians of the island that would still puzzle man down to the 21st century. The natives exhibited considerable curiosity about the Dutchmen. By morning, a large number of natives had gathered along the shore. Somehow, a shot was fired, scattering the crowd. When the natives returned, Roggewein, at the head of 150 men, fired a volley into the crowd. Terrified, the natives offered everything they had to appease the powerful and violent men who had reached this island from afar. The giant statues had done little to protect them when the Dutchmen touched their shores.

Eventually a violent windstorm forced the ships to leave their anchorage, and they continued west-northwest. There was little to be discovered in the area.

In fact, most of this area had been previously explored. Roggewein passed an island which he believed to be the Island of Dogs, discovered by Schouten. For another day, they were driven by wind and current, until they came to a group of low islands. The African ran aground on a coral reef on April 19. She was lost, but all hands were saved and divided between the Eagle and the Tienhoven. A short time later, the Tienhoven nearly met the same fate on a low lying island which Roggewein named Aurora. Roggewein continued sailing at about 15 or 16 degrees, when he suddenly found himself in the midst of a group of half submerged islands. Fortunately, it was calm enough for them to slowly extricate themselves. They found a beautiful island paradise with much needed coconut palms. Roggewein sent an armed detachment ashore. Once again, the Dutchmen showed their penchant for not establishing good foreign relations, by firing into the crowd of curious onlookers. Trying to undo the damage, they switched tactics and offered presents to the chiefs. The natives were no longer in a friendly mood, however; and managed to draw a group of Dutchmen into an ambush. Ignoring the superior firepower of Roggewein's soldiers, the natives were able to inflict several casualties and force the withdrawal of the Europeans.

In spite of the struggles, the men were refreshed by their stay at this island and named it Recreation Island. Roggewein extensively searched the Pacific, suffering heavy losses in men through illness. Sometime after leaving Recreation Island, a council was held and it was decided to head the two ships north to the East Indies. At most stops, Roggewein tended to fight with the natives. The vessels touched on New Ireland, and New Guinea, and crossed the Moluccas to Batavia where they cast anchor.

Batavia was a stronghold of the Dutch East India Company and the governing officials arrested Roggewein and his crews and confiscated the Eagle and Tienhoven.

Later, after a trial, the East India Company was forced to restore what it had taken and to pay heavy damages. By being sent home prematurely, however, Roggewein reached Texel on July 11, 1723, just one year, ten months, and 20 days after his departure, making his circumnavigation the fastest in history, and the first to break the two year barrier. Forty-three years later, this record would be broken by an Englishman, but only by a scant two days.

Alan Boone

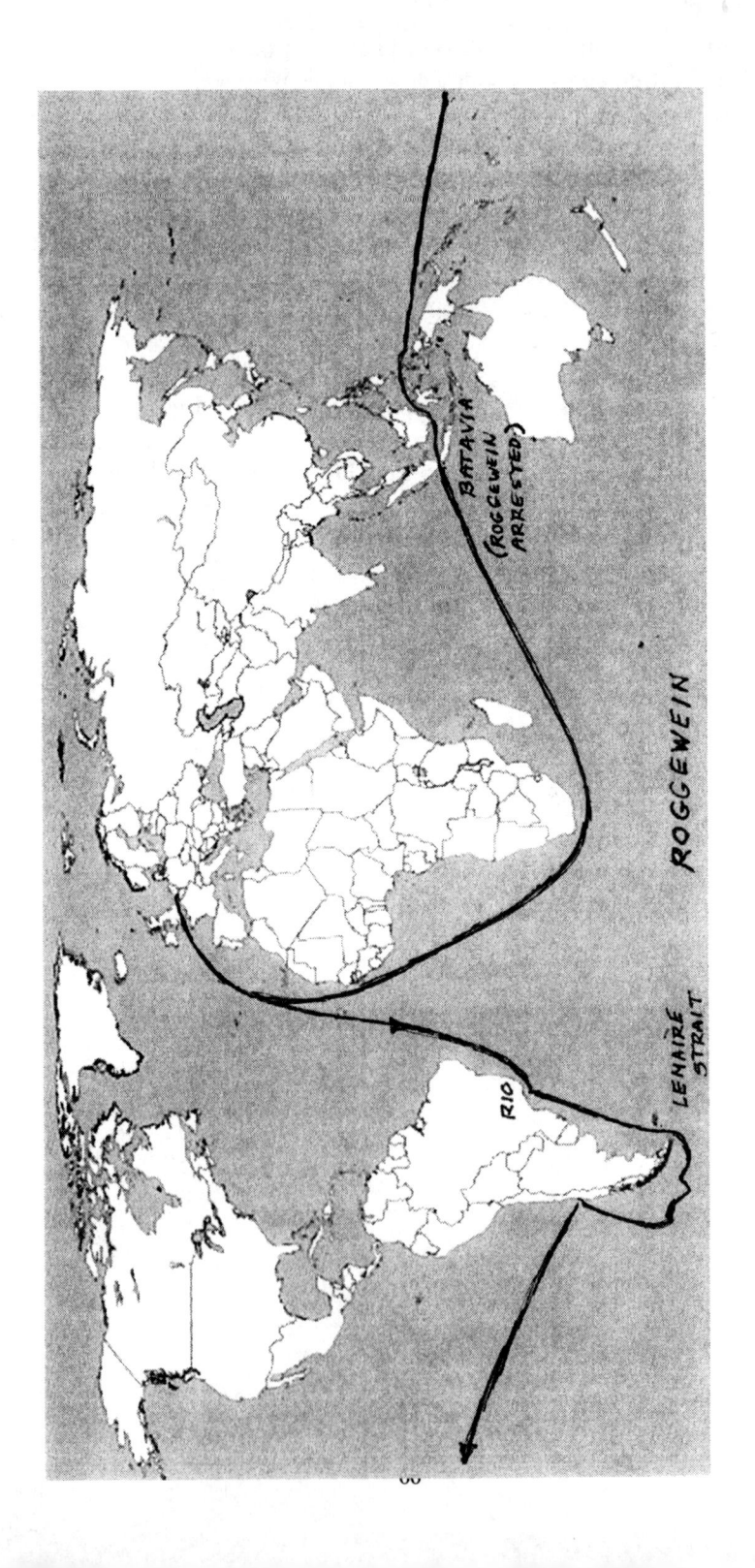

CHAPTER 6 - BYRON

Sailing ships had changed considerably between the time of Schouten--LeMaire's voyage and Roggewein's voyage. Gone were the high poop decks and forecastles. Ships had become larger, longer, lower, and carried more sail. They were faster and more seaworthy. Later, copper sheathing was added to the hull to help the ships glide through the water. The year that Roggewein returned to Texel, a man was born who would break Roggewein's record in a copper sheathed vessel. His name was John Byron, and he was an Englishman.

Byron was born of nobility, the grandson of Lord Byron. He went to sea at the age of 17 and sailed on Admiral George Anson's voyage around the world in 1740, a military expedition against the Spanish on the Pacific Coast of South America. Byron had been on the Wager which was wrecked in the Strait of Magellan. The crew, including Byron, had been taken captive by the Spanish. After being a prisoner for three years, Byron managed to escape, being rescued by a vessel which returned him to Europe. After his return, he immediately went back into the service and later distinguished himself in the war with France.

Seventeen sixty-four was a time of relative peace for England, whose empire was beginning to take shape. On June 17th of that year, Commodore Byron received his orders from the Lord of the Admiralty.

"As nothing contributes more to the glory of this nation, in its character of a maritime power, to the dignity of the British crown, and to the progress of its national commerce and navigation, than the discovery of new regions; and as there is every reason for believing in the existence of lands and islands in great numbers, between the Cape of Good Hope and the Strait of Magellan, which have been hitherto unknown to the European powers, and which are situated in latitudes suitable for navigation, and in climates productive of different marketable commodities; and as moreover, His Majesty's islands, called Pepys and Falkland Islands, situated as will be described, have not been sufficiently examined for a just appreciation of their shores and productions, although they were discovered by English navigators; his Majesty, taking all these considerations into account, and conceiving the existing state of profound peace now enjoyed by his subjects especially suitable for such an undertaking, has decided to put it into execution."

In addition to the wordy orders, Byron was given two ships. An additional supply ship would be sent to reprovision him in South America and return. The two ships in the fleet were the Dauphin, a sixth rate man-of-war with 24 guns and a crew of 190 men, and the Tamar, a sloop of 16 guns, and a crew of 120 men. Byron placed Captain Mouat in command of the Tamar. On June 21, they left the Downs, and immediately the Dauphin grounded in the Thames River, and had to put in at Plymouth for repairs. Following this inauspicious beginning, they sailed from Plymouth on July 3, and, after a journey of ten days, anchored at Funchal on the island of Madeira. Later, Byron put in at the Cape Verde Islands to take on fresh water. The two ships crossed the Atlantic in tropical heat and constant rain, which caused a considerable amount of sickness in the crews. They looked forward to a rest in Rio. During the trip, Byron noted that the copper sheathing seemed to disperse the fish from the ship's hull.

At Rio, Byron was welcomed to the Portuguese city with much pomp and a fifteen gun salute from the nearby fort. Byron was greeted at the palace by his Excellency and, after a brief meeting, was conducted ceremoniously back to the ship. The sick were disembarked and cared for and the ships were restocked.

The tropical heat continued to be unbearable to the Englishmen, and they eagerly weighed anchor on October 16 to head south to cooler climates.

Now, Byron was able to announce to the crews the secret nature of their mission. To offset the unhappy prospect of a long voyage of exploration, Byron was able to proclaim that the Admiralty had approved double pay and advancements to those sailors who served well on the voyage.

On October 29, a violent storm struck the two ships and the intensity increased until Byron had to order four guns thrown overboard to keep them from foundering. In November, the stormy southern seas provided an illusion of land for the sailors. Amid the thunder and lightning, the mountains of a ghostly shoreline appeared and remained. Only the sudden clearing of the weather revealed there was nothing but sea and sky. Byron declared that in all his years at sea, he had never seen such a complete and sustained illusion. They were nearly 44° south and 60° west.

The next day, a powerful gale struck, tearing the mizzen mast sail to ribbons. Fighting storms and gales, they managed to reach Penguin Islands and Port Desire on November 24. Port Desire had never been the paradise that the name implied. It was barren and cold, and although there were sufficient seals and birds to feed the crews, only poor, brackish water could be found. Byron quickly sailed for Pepys Island in accordance with his orders. No one was quite certain of the position of this island. Byron found nothing in his search except storms and waves higher than anything he had witnessed before.

He returned to South America at the entrance to the Strait of Magellan. Byron was determined to see the now legendary Patagonians. He landed with an armed contingent of soldiers. They too, found natives of gigantic stature, naked except for skins and brightly painted decorations on their bodies.

At some time in the past two centuries, horses had been introduced and many of the natives rode horses along the slippery rocks at the shoreline. Byron distributed gifts and found the natives to be friendly and responsive.

Byron entered the Strait of Magellan in search of a place where they could find wood and water before beginning a search for the Falkland Islands, which lie to the east of the straights. He found shelter at Port Famine, where Cavendish had found the starving survivors of the Spanish garrison. There was plentiful wood floating near the shore and many birds in the trees. They stayed at Port Famine for nine days, until the crews were rested and the ships were restocked. Then, they set out in search of the Falklands. Upon reaching those islands, they found an excellent harbor which Byron named Port Egmont in honor of the First Lord of the Admiralty. In this generally barren land, they found wild sorrel and celery, both known to have value in preventing scurvy. There were geese, seals, walruses, and a foxlike animal of unknown origin on this island, surrounded by a great expanse of water.

After exploring extensively the Falklands were left and Byron returned to Port Famine. He met with the Florida, the supply ship sent from England, on January 27. Reprovisioned by this meeting, he was ready for the next leg of his journey. Byron had been around Cape Horn, formerly Cape Hoorn, and felt there were many advantages to passing through the Strait. The hazards of the Strait came only when attempted at the wrong time of year, and if traversed properly offered a chance to collect many anti-scurvy fruits, grasses, and vegetables. Byron's delay on this trip, which had taken from February 17, to April 8, 1765, had been because of the stormy equinoctial season. He entered the Strait in order to expedite his journey.

Many times, while in the Strait, they were visited by the Patagonians. The Englishmen were appalled by the wretched state of the natives. They lived on mussels, some fruits, and putrid fish thrown up on shore. They were observed tearing at the carcass of a dead whale with their teeth. The Strait was not a friendly area whether on land or sea. Also in the Strait, they met a French ship, the Aegle, searching for wood. Even in this barren and forlorn area, shipping was becoming common.

By April 26, they had reached Juan Fernandez Island, Selkirk's lonely abode. This had become an essential stopping place for ships traveling in these seas.

Byron sent men ashore to get wood and water, and to catch some wild goats for meat. When it was time to return from the foraging expedition, the ship's boat was unable to get close to shore because of the rough surf. One sailor refused to try and reach the boat, since he couldn't swim. The other men threatened to maroon him Selkirk fashion. They gave him a life jacket, but he still refused to move. His companions tied a rope around him and dragged him from the shore with the boat. Upon reaching the ship, he appeared lifeless, and the crew hung him upside down until he regained consciousness. His recovery was complete and the expedition avoided a casualty.

Leaving Juan Fernandez, Byron searched in vain for Easter Island. After spending eight days in fruitless pursuit, he headed northwest to find the

Solomons. On May 22, scurvy broke out and rapidly spread through the crews. They spotted two islands with an abundance of coconuts and tropical fruits, but were unable to anchor. The sick men could only look on as they passed these meccas of lost hope on June 8. In the agony of the moment, they charted them, naming them Disappointment Islands. The next day, they found another island covered with coconut palms. As the boat approached shore the crewmen were attacked by natives. Seeing that the English were outnumbered, Byron had them dispersed with gunfire, killing three or four of them. This was sufficient to keep the natives at a distance so the crew could gather coconuts and other plants. The fresh fruits were so effective in overcoming the scurvy that within a few days, all of the sick had been cured.

The ships were soon picking their way through the islands in the western Pacific near the equator. Dysentery plagued them in the oppressive heat, and only putrid, brackish water was left to drink. On July 28, Byron reached Saipan and Tinian in the Marianas. In the steamy jungle islands, the crew set up tents for those who had redeveloped scurvy after the fruits and plants previously foraged had been consumed. The heat continued and malaria fever was added to their woes. Although many tropical fruits could provide relief from the scurvy, two sailors died from the fever. After nine weeks of recuperation, and the restocking of the Dauphin and the Tamar, they left Tinian and sailed on a northerly course before turning south and reaching Sumatra on November 20.

Byron cast anchor in the harbor of Batavia, the Dutch capital in the East Indies. A polyglot population had built a prosperous trading city at Batavia with fine well-kept canals, pine bordered streets, and sturdy buildings. The English crews enjoyed the civilized reception received at this city, but had to shorten their stay out of fear of the deadly fevers indigenous to the area. The English fared better in the Dutch colony than either Shouten-LeMaire or Roggewein, sailing with their crews without intervention. However, no sooner had they left harbor, than a fever broke out and spread through the ships, causing the death of three sailors.

The ships crossed the Indian Ocean in 48 days and cast anchor at Cape Town, quickly loading provisions and turning for home. Sailing north in the Atlantic the Tamar received a violent shock, as if she had struck a rock. On rushing forward, the men saw the sea red with blood, reminiscent of the collision between Schouten's Eendracht and a great fish. Later, it was discovered that the Tamar was so dilapidated and her rudder so damaged that she could not continue the voyage. The Dauphin persevered, dropping anchor on May 9, 1766, having circled the globe in an unprofitable, non-fulfilling, albeit record breaking time of one year, ten months, and 18 days. The expedition was not made to bring home a cargo of spices, gold, or pirated booty. It was a voyage of discovery, which discovered little. It was also meant to display the English flag around the world, which the Dauphin accomplished. Her record would stand for 26 years, until wrested from the English by an enterprising Frenchman with the romantic moniker of Etienne Marchand.

Circumnavigation

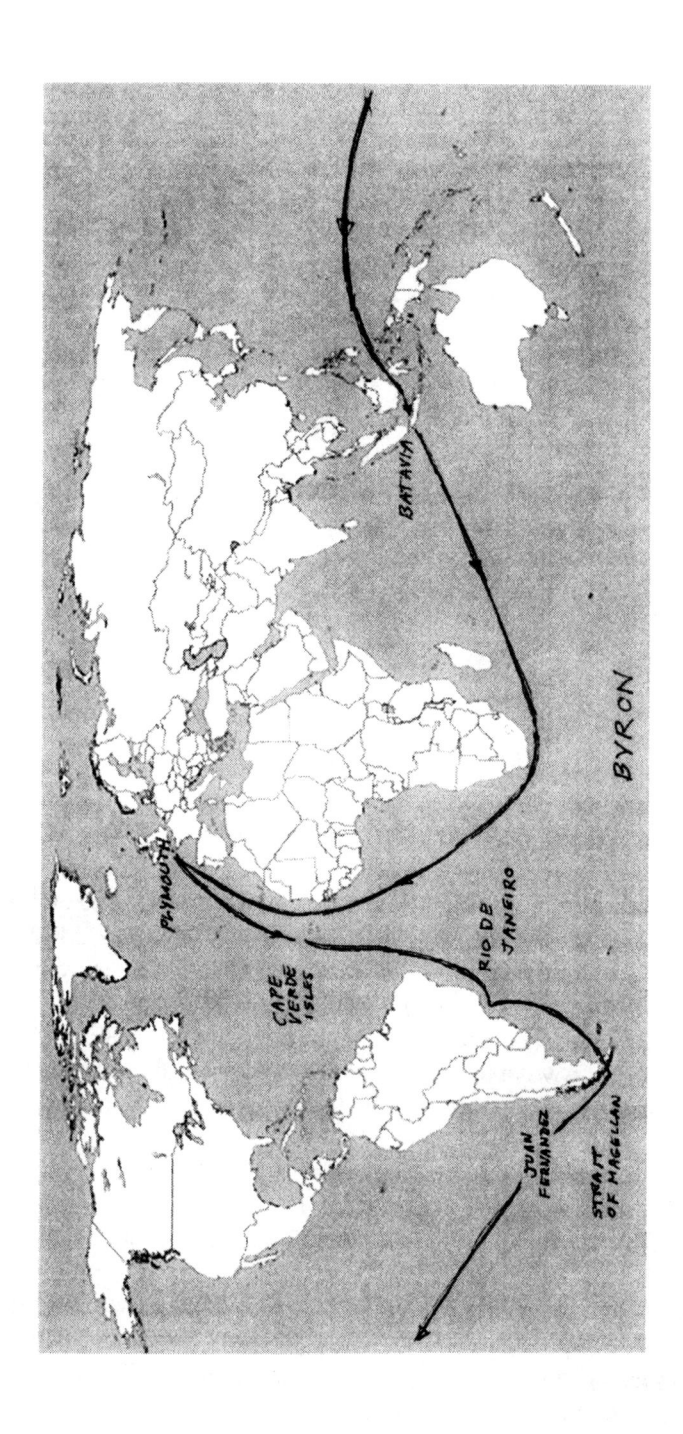

CHAPTER 7 - MARCHAND

The world of the late eighteenth century was in the throes of change. History had passed through the age of exploration in which much was learned of the face of the planet. During the Renaissance there had been renewed interest in art, literature, and learning. Science, technology, and the industrial revolution were changing societies. The eighteenth century was also known for its great philosophers. Perhaps this is why, near its end, it was marked by social revolution. The obstreperous citizens in the American colonies began a war of independence in 1776, which lasted until 1789. The French Revolution began in 1789, and lasted for ten years. The voyage of Marchand took place entirely during the French Revolution, an opportune time to absent oneself from that land.

"There are furs on the Northwest coast of America, furs that could be purchased from the Indians for a mere pittance in trade goods," the English sea captain said to Etienne Marchand at St. Helena in 1788.

Marchand, returning from a voyage to Bengal, made careful note of what the man said. If these furs could be taken to China, they would yield a handsome profit. A cargo of Chinese imports carried to Europe would enhance that profit even more.

Marchand returned to France and enlisted the aid of M. M. Baux, an entrepreneur who financed the venture of Marchand's dream. This circumnavigation would be in contrast to Byron's government financed venture. Marchand was going for private profit only.

Baux and Marchand decided to act quickly, and began construction of a 300 ton vessel. This single ship represented greater tonnage than Francis Drake's entire fleet and was the first time a record circumnavigation would be undertaken with a single ship.

Marchand knew that sailing the Pacific called for a sturdy craft, capable of being stocked for three or four years at sea and with proper armaments. They named the ship the Solide. She was copper plated and modern in every possible way. The crew would consist of Marchand, two other captains, three lieutenants, two surgeons, three volunteers, and 39 seamen, a total of 50. The armament was four cannon, two howitzers, four swivel guns, and an ample supply of powder and shot. Her hold was stocked with trade goods and repair parts.

Previous experiences with long ocean voyages had been studied, and this information, which had not been available to Magellan and his counterparts, would serve to make the ship comfortably habitable. Careful attention was paid to the future health of the officers and men. Proper foods and sanitary facilities would make this one of the safest voyages of the century.

Marchand had hoped to be at Cape Horn by the beginning of the northern hemisphere winter, but the Solide was not ready to leave Marseilles until December 14, 1790. They sailed through the Strait of Gibraltar and into the Atlantic. Marchand stopped for a brief stay at Praya in the Cape Verde Islands before heading directly for Staten Island at Cape Horn, which he reached on April 1, 1791. He had been at sea only three and one half months. He then passed around Cape Horn and headed northwest.

Because of winds and currents, the Solide headed far into the Pacific, thus avoiding the calms which characterized the more eastern part of this ocean. It was Marchand's desire to cross the equator at about 142° west longitude and head directly toward the northwest coast of America. By early May, however, the drinking water began to become brackish and he decided to head for the Marquesas Islands. By this time, most of the Pacific had been explored by such people as Captain Cook, and the

Solide was traversing known areas. The Marquesas had been discovered by Mendoza in 1595, and visited by Cook in 1774. Constant astronomical observations allowed the Solide to maintain her course during the 73 days after leaving Staten Island. During this time, no land was sighted.

On June 12, 1791, they sighted Magdalena Island, the southernmost island in the chain. The Solide continued north through the chain until she anchored in the Madre de Dios Bay where they were welcomed by friendly islanders. In this proximity to the equator, about 6° north, the tall brown natives had little need of clothing. They were fashion conscious, tattooing themselves, piercing their ears, and wearing bright necklaces of red beads and lightweight wood. The Frenchmen found this picturesque island paradise friendly and delightful. The natives enjoyed quantities of brandy, but never showed signs of drunkenness. The island's housing consisted of huts placed on intricately carved stilts. The islanders worked at building canoes and weapons, fishing, singing, and dancing. It was an easygoing, relaxed existence. The Frenchmen did not avail themselves of one friendly custom of the people, that of chewing food and offering the pre-chewed morsel to a friend so that he need only swallow.

Marchand kept noticing a spot on the horizon each clear evening at sunset. When he finally left the known islands, he sailed north toward the spot, and thus became the discoverer of several unknown islands north of the Marquesas. On leaving his discovery, Marchand headed for America, sighting Engano Cape on August 7. The Solide had been at sea for 242 days, and had spent only ten days in port. There had not been a single case of scurvy during that time, a tribute to progress, and to the captaincy of the voyage.

Engano Cape was called the port Tchinkitane by the natives. Native names were mostly, if not always, ignored by new "discoverers" who renamed them with impunity for queens, countries, or whims.

In port, Marchand was able to buy high quality otter skins. The Frenchmen found these natives to be less attractive to them than the people in the Marquesas. Their faces were covered with a thick coating of grease, and the women inserted wood into a slit in their lower lips.

The Solide left Tchinkitane on August 21, 1791, and steered southeast in search of Queen Charlotte Island, off the coast of British Columbia. They quickly entered Cox Strait and traded with the Indians for furs. The low, rocky shore was covered with fir trees. Searching for better trading, Captain Chanel explored thoroughly to the south. As the Solide reached Berkeley Bay, a ship of three masts was seen approaching from the south. Marchand knew that the key to bringing a good price in the world markets was to be the first to market with choice goods. He determined to leave immediately for China, via the Sandwich (Hawaiian) Islands, discovered earlier by Captain Cook.

After a fast passage, the Solide reached Hawaii on October 5. Marchand carried on trading aboard the Solide. Here, he obtained pigs, coconuts, bananas, pumpkins, and watermelons. The latter two had apparently been developed from seeds sown by Cook during his voyage. Marchand never tarried long in port and after four days, the Solide sailed on toward China. They passed by Tinian in the Marianas, and headed for the southern part of Formosa. Passing this, they reached Macao on November 28, 1791, less than a year after leaving Marseilles.

Quick profit eluded Marchand. The Chinese government had banned the importation of furs and set heavy penalties for violation of the law. Nobody knew why. They only knew that they could not unload their cargo here. No one understood the Chinese reasoning, but the mandate was clear.

Marchand wrote to Baux's agents in Canton to see if he could sell his cargo there, but the prohibition covered all of the Celestial Kingdom. He could have sold his cargo at Whampoa, but the taxes would have taken such a bite out of the profits that this was impractical. Marchand could only return to France to market his furs. It would not have been characteristic for Marchand to waste any time, and he sailed for Mauritius, once the decision became clear. Rounding the Cape of Good Hope, the Solide passed through the Strait of Gibraltar, and reached Marseilles on August 14, 1792, just one year and eight months after leaving. He had circled the globe and travelled up and down two oceans. With the fast passage of the Solide, an era came to an end. It would be nearly eighty years before the former water route around Cape Horn and the Cape of Good Hope would have a viable alternative. All of these first seven record breaking circumnavigators had travelled from east to west by this route. This passage would be the quickest way until the opening of the Suez Canal and the Panama railway many years later, although later sailings were to go from west to east.

Great men in slow, sturdy sailing ships had made the world smaller by reducing the trip from three years to one and two thirds years in less than three centuries. No longer was an appalling loss of life and ships necessary to complete the journey. Although the route would not change for many years, technology would. The names of great men were soon to be replaced with the names of great ships in the coming era. At the time of Marchand's journey, the steam engine and steam powered shipping was just developing, but it would be the next generation of sailing ships which would shrink the globe by nearly two thirds in a short space of time. Technology would evolve on the drawing boards and in smaller waters, while sailing continued to flourish. Even with the rapid changes, Marchand's record would stand for over half a century.

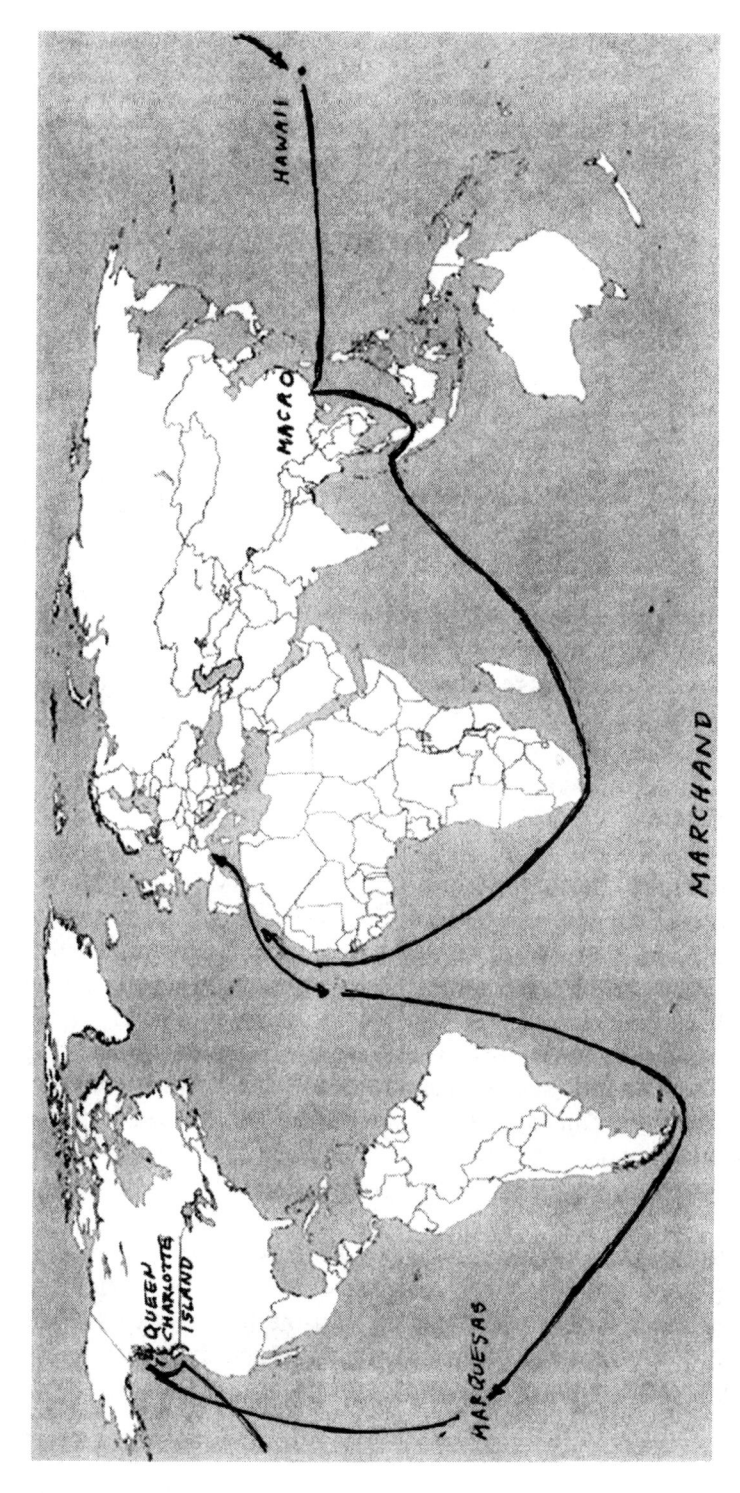

Circumnavigation

CHAPTER 8 - THE CLIPPERS

Following the voyage of Marchand and similar voyages by the American Robert Gray at about the same time, several circumnavigations took place at a much more leisurely pace. It was failure that added speed to some at the early voyages, cutting them short. Willem Schouten and Jacob Roggewein had their voyages cut short by circumstances and Etiene Marchand was unable to trade in China as he would have preferred.

The exploration of the world was not yet complete, and the English and French conducted scientific voyages well into the nineteenth century that completed circuits around the globe. The Russians tied the northeast segment of their empire to Northern Europe with several notable voyages between 1803 and 1829. Charles Darwin completed his lengthy scientific voyage in 1836. The American, Matthew Maury completed one of the most fruitful circumnavigations in 1830. His studies on winds and currents, combined with the advancing technology of sail and steam powered craft, was soon to create a drastic reduction in the size of the world in the middle of the eighteen hundreds.

Leading to the shrinking of the world was the new desire to sail around the world quickly. Until now, speed had not been a factor, and record breakers had done so accidently. But the new treasures of silk, opium, tea, and gold would change the impetus, and drive men to faster, profit conscious circumnavigations.

About the same time Marchand was planning his voyage, American John

Fitch was starting the first steam powered commercial passenger and freight service. In the summer of 1790, he made scheduled runs up and down the Delaware River between Philadelphia and Trenton. Developed by a Scotsman in 1769, the steam engine was first adapted to a boat by a Frenchman in 1783, and made workable by an American in 1787.

Robert Fulton made the steamboat a money making proposition in 1807, and in 1809, a coastal voyage from New York to Philadelphia was made by the steamship Phoenix. In 1819, the Savannah crossed the Atlantic Ocean using up its supply of fuel and finishing the voyage under sail. In 1838, the side wheeler Sirius crossed the Atlantic under steam power alone. Steam had a drawback, though. It required fuel. Fuel, such as coal or wood, required space. Space devoted to fuel could not be used for cargo, and the less cargo, the less the profit. Moreover, naval architecture was producing sailing vessels of ever greater speed. For a while, steam would be an idea whose time had not yet come, even though the technology existed.

Instead of waiting for full cargoes, and then sailing, the Ball Line instituted scheduled service between America and Europe in 1818. This became so popular that soon scheduled packet ships were plying back and forth between the two continents regularly. The ships were small and sturdy and carried passengers in comfort heretofore unheard of.

In faraway China, though, lay the key to big profits. Tea, whose quality deteriorated during a long sea voyage, was a commodity that was very much desired in the markets of London and New York. The sensitive palates of these countries paid premium prices for the cargo that spent the least time at sea. The route to China took ships east from New York or London, around the Cape of Good Hope to Chinese ports then back over the same passage. It was this trade that first spurred the development of fine lined sailing ships known as "clippers," because they "clipped" off the miles. By 1846, the 757 ton Rainbow had completed this voyage in a little over six months. Small, fast ships designed for "speed to China" were not built specifically for the rough sailing around Cape Horn and in the Pacific, and there was no reason for Americans and their rivals, the British, to circle the globe for their tea or opium.

However, a ship carrying a cargo from New York around Cape Horn to Valparaiso, Chile, was already far on its way to the Celestial Kingdom of China and could cross the Pacific to Canton and return to New York around the Cape of Good Hope, with her cargo of unspoiled tea rapidly reaching port. Such was the case with one of the first ships to be built to carry the name clipper.

She was the Ann McKim, a yacht-like 193 ton ship built in Baltimore in 1832, for the Far East trade. On her first voyage to China, the Ann McKim sailed around Cape Horn to Valparaiso on the west coast of Chile, thence across the Pacific to Canton, making the return voyage west from Canton to New York in 150 days. Her route was shorter than Marchand's, but still qualified as a trip around the world.

The Ann McKim was captained by George Vasimer. She left New York for her record breaking trip on December 18, 1840, returning on December 8, 1841. Her trip constituted a voyage around the world in just ten days shy of one year. Half a century after Marchand, the speed had been cut by eight months. It should be noted that the Vasimer was interested in speed, and the trip by Marchand in the Solide would have taken even longer than it did had his wishes to trade in China not been denied. There is no way to know how long the Solide would have taken to make the circumnavigation using Vasimer's route and desire for speed. Of such enigmatic pieces is the picture of history composed, with the final recording of facts the only true chronicle.

The race for China tea was just beginning. Great ships were being built for the run to China and back, fast ships with sharp clean lines and carrying lots of sail. They had names like Hougua, Rainbow, Samuel Russel, Memnon, Oriental, and Architect, all built between 1844 and 1849.

The fastest ship of her day was the Sea Witch. She set the first sailing record which still stands today. Launched December 8, 1846, she was hailed by the New York Herald as "the prettiest vessel we have ever seen." The 908 ton ship was the tallest ship afloat. Her hull was painted black with a gilt stripe and her figurehead was a Chinese dragon. During her ten year career, before she was wrecked off Cuba on March 26, 1856, she proceeded to break one sailing record after another. The Sea Witch made two fast passages to Canton under the command of the hard driving Captain "Bob" Waterman; returning the same way to New York.

On her third voyage, the Sea Witch followed the course of the Ann McKim. She left New York on April 25, and bucked the strong westerlies around Cape Horn to Valparaiso in 69 days. She sailed from Callao to Hong Kong in 50 more sea days. Her homeward journey from Hong Kong to New York was 77 days for a total sailing time of six and one half months. When she arrived at New York on March 25, 1849, Capt. Waterman had taken her completely around the world in eleven months, cutting one month from the record of the Ann McKim.

On her fourth voyage, the Sea Witch repeated her feat over the same route. She left New York on April 12, 1849. She cut one day from the sailing time from New York to Hong Kong via Valparaiso to 118 days, and ran home in 74 and one half days, arriving March 7, 1850. After this trip there were new records under her belt including the tenth record for speed around the world.

That record was ten months and 23 days.

Gold was discovered at Sutter's Mill, California in 1848, and suddenly a market opened up for fast ships to carry gold seekers and their supplies to the turgid economy of San Francisco. The great race of the clippers around Cape Horn to San Francisco was underway, and speed was measured in terms of sea days between New York or Boston and San Francisco. California clippers sailed on to China and circled the world by returning westward to New York.

Another gold strike followed fast on the heels of the California strike, this time in far off Australia, the Terra Australis which had been searched for by so many of the early circumnavigators. The year was 1851. This spurred the development of fast seagoing tonnage as hundreds of immigrants sailed in search of their pot of gold at the antipodes.

A new, shorter route around the world evolved from these circumstances. Fast ships leaving Liverpool sailed directly south through the North and South Atlantic until they reached the "roaring forties." In the regions of 40° south latitude, strong westerly winds roared in a seemingly continuous pattern. Ships would unfurl their sails before these winds and be quickly driven to the east. The same winds that hindered the early Circumnavigators, and turned back the Bounty, became the ally of these sailors as they sped to Melbourne, past the Cape of Good Hope, picked up a cargo and sailed on eastward past Cape Horn. One such ship was the Phoenician. This fast Scottish clipper ship soon developed a reputation for speed. Owned by the Aberdeen White Star Lines, and under the command of Captain Sproat, she made three extraordinarily good voyages between London and Sydney, bringing the around the world time down to approximately six and one half months. Her first three voyages were:

	London to Sydney	Sydney to London	Total Sea Days
1849-1850	90 days	88 days	178 days
1850-1851	96 days	103 days	199 days
1851-1852	90 days	83 days	173 days

Allowing two weeks in port, her total times for the first and third voyages should have been six months, twenty days and six months, fifteen days, respectively.

The 526 ton clipper had twice sped around the globe in under seven months, but her time was soon to be beaten.

James Baines of Liverpool started the British Black Ball Line in 1851, to begin a scheduled packet service from Liverpool to Melbourne. When Baines purchased the Marco Polo, a Canadian ship built in 1851, he was able to buy her for a low price. English critics had described her as being square as a brick fore and aft, and with a bow like a savage bulldog. Baines knew a good ship when he saw her and the criticism only helped drive the price down on a vessel he believed in.

The Marco Polo was a three deck ship of 1,625 tons, 185 feet long with a 38 foot beam and a 30 foot deep hold. She had a flat deck without poop or forecastle, unlike the earlier sailing ships. She was well timbered and sturdy, and was to remain in service until being wrecked in 1883.

On her maiden voyage to Australia, under the command of Captain "Bully" Forbes, she sailed from Liverpool on July 4, 1852. Captain Forbes was another hard driving clipper ship captain, with a reputation for padlocking his sheets and carrying pistols to shoot at any crewman who was not hurrying to his satisfaction. The Marco Polo, hailed as a slow, sluggish mistake, reached Australia with her load of passengers a week ahead of the rival steamer Australian. She had followed Matthew Maury's recommended course.

Once in port, Forbes faced the prospect of desertion. Many crewmen tended to wander off to the gold fields of Australia, much as they did in California. Forbes solved this by confining his crew to the ship while in Melbourne taking on cargo. After 26 days in port, she sailed for home, breaking speed records along the way. As cargo, she carried a few passengers, $100,000 in gold dust and a 340 ounce gold nugget purchased by the British government for the Queen. On November 3, she made her best day's run of 353 miles at an average speed of 14.7 knots per hour.

On December 26, 1852, she returned home, having sailed around the world in the record breaking time of five months and 21 days. As she sailed into port, a large strip of canvas hung between her fore and mainmast declaring her to be "The Fastest Ship in the World." Suddenly the world took notice of the ship which had been dubbed as a scow. Prior to this second run in the Marco Polo, Captain Forbes claimed, "Last trip I astonished the world with the sailing of this ship. This trip I intend to astonish God Almighty." Although she remained an impressive vessel, the Marco Polo would not equal her own record again. Her second voyage was made in six months even.

Even in the fast breaking decade of the fifties, the outstanding performances of the Marco Polo remained the record for nearly two years. Another fast ship, intended for the Australian trade, was built at Rockland, Maine, by George Thomas, and launched on November 2, 1853. She was the 2,305 ton Red Jacket, one of the speediest of all clipper ships. She was 251 feet long with a 44 foot beam. She was purchased by the White Star Line, rival to James Baines and his Black Ball Line. She was said to have a delicate and graceful beauty. On her maiden voyage, she sailed from New York on January 10, 1854, and crossed to Liverpool in just 13 days, one hour, 25 minutes dock to dock. With speed becoming more and more of a selling point, times were being kept by the clock, instead of the year, month, or even day.

Once chartered for Australia, Captain Samuel Reid took command and sailed from Liverpool on May 4, 1854. Ten days later, the famous Captain James Nicol "Bully" Forbes took the new Donald McKay clipper Lightning out of the same port with the same destination. The 2,084 ton Lightning had been built at East Boston and launched January 3, 1854. Sailing for the Black Ball Line, Forbes boasted that he would be in port before the Red Jacket. A race around the world was on, and before it was finished, the record time was to fall twice in eight days.

The Red Jacket averaged 8.6 knots per hour for over 13,880 miles. She passed the Mermaid, which had sailed the day before, off of the coast of Spain. After passing the Cape of Good Hope on June 24, the Red Jacket picked up strong winds. Sailing as far south as 52°, she met gales mixed with sleet, hail, and freezing temperatures. She made good time, covering 400 miles on one windy day. On July 12, the Red Jacket arrived at Melbourne, after a passage of 69 days. On July 31, 78 days after sailing, the Lightning entered port. Big mouthed "Bully" Forbes' boast had not come to pass. He became even more determined to win the race home.

The Red Jacket was in Melbourne 22 days compared to the Lightning's 20 day pit stop. Speed was important insofar as it contributed to profits for the sailing lines, and business had to be conducted boast or no boast.

On August 3, the Red Jacket left Melbourne. Australian authorities, thinking Captain Reid's White Star representative was going to abscond with her gold dust cargo, sent two warships after him, but they were unable to catch the Red Jacket.

The ship was poorly loaded. It was too light overall and too heavy in the stern. However, she made a fast trip across the South Pacific toward Cape Horn. Off Cape Horn, she encountered ice floes and ice bergs and had to dodge them. At night, she had to furl sails completely and do no sailing in the treacherous dark waters. For three days, she was imprisoned by the frigid conditions and hopes of a fast passage seemed to be dashed like a light vessel against the heavy bergs.

Once clear of the ice, the Red Jacket made up the lost time, doing well against contrary winds, finally reaching Liverpool on October 15, 1854. She had made the passage in 72 1/2 days, and had completed a voyage around the world in five months, ten days, 22 ½ hours, and knocking ten days from the Marco Polo's record. She had carried nearly a million dollars' worth of gold dust into port. Her glory would be short lived, for the race was continuing.

Having left Melbourne on August 20, 1854, Captain "Bully" Forbes had been driving the Lightning relentlessly 24 hours a day. His stubbornness had lost him many topsails in the chase, but he kept to his goal of speed. When he arrived at Liverpool on October 23, 1854 he had completed the circumnavigation in five months, eight days, and 21 hours. His arrival was just eight days after the Red Jacket and shaved one day off the record. Forbes was a winner, but his record would stand for less than a year.

One more record was to climax the clipper ship era. That was a performance of the James Baines that still stands among sailing records. The 2,525 ton James Baines was another of the large, fast clippers built by Donald McKay for the British-Australian trade.

The James Baines started her career by breaking the transatlantic crossing record by sailing from Boston to Liverpool in 12 days and six hours.

On December 10, 1854, Captain McDonnell sailed the James Baines from Liverpool. She was held back by strong headwinds and did not clear land until December 16. Her cargo consisted of 250 sacks of mail, containing over 180,000 letters and newspapers, 80 first class passengers, 622 immigrants, a bullock, 75 sheep, 86 pigs, and 1,200 fowls and ducks.

Under a contract that the Black Ball Line had with the British government, a 100 pound per day penalty would have to be forfeited for each day over 65 days for the journey to Melbourne. Even though the James Baines struggled against light northerly winds during her journey southward in the Atlantic, she picked up fair winds in the "roaring forties" and reached Melbourne in 63 days, 18 hours, and 15 minutes. The date was February 12, 1855, and a pleased McDonnell had no penalty to pay, and a new world record for speed.

She stayed in port until March 12. She completed the return trip in 69 1/2 days. The maiden voyage of the James Baines had set a record which would stand for an amazing 35 years.

Her arrival at Liverpool on May 21, 1855, was five months and eleven days after she left on December 10, 1854. These were short months, however, and with 134 days at sea and 28 days in port, a total of 162 days passed between the date of departure and the date of arrival. This compares to the Lightning 163 day run between May 14 and October 23 1854, and the Red Jacket's 164 day run from May 4 to October 15, 1854. Like all record holders since the Marco Polo, she had gained a day by travelling from west to east. The earlier circumnavigators had lost a day by sailing from east to west.

The clipper ship era was soon to end. Changing economic conditions would cause the need for fast sailing ships to dissipate. Some of these once proud vessels would lie rotting in San Francisco harbor for want of companies and crews. Others would be relegated to carrying coal or guano on short runs.

As the ships became older, their speed slowed and it was soon no longer the current state of the art to build fast sailing ships. The final death knell of the sailing ship came when the Suez Canal was opened in 1869, cutting a vast amount of time from the trip east. Sailing ships were at a distinct disadvantage against the steamers. Even if they could be towed through the canal, they could be lost in the tricky winds of the Arabian Gulf.

Again, there was no longer a reason to go around the world. Travel to India, China, and Australia could be quickly accomplished by a short two way steamer trip through the canal. The record would not fall again until a motive was established for circumnavigating our orb. This impetus would come, strangely enough, from a French writer of fiction.

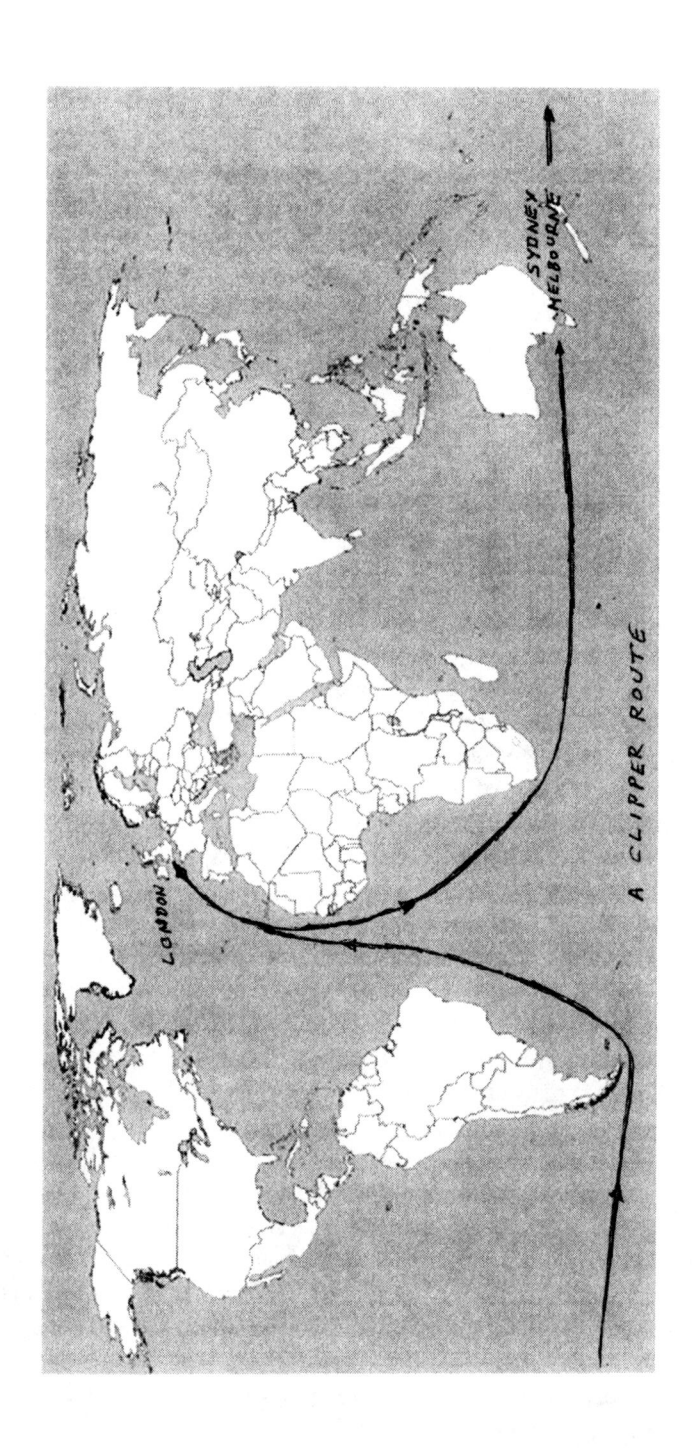

Circumnavigation

CHAPTER 9 - COCHRANE

It was in 1872, according to the fictional account of Jules Verne, that Phineas Fogg, an English adventurer, and his faithful servant Passepartout traveled around the world to win a wager. Verne's book, *Around the World in Eighty Days*, was as much in the realm of fantasy as his travels under the sea or his voyage to the moon tales. The unemotional character of Fogg mapped out a route from London to Suez by rail and steamboat in seven days, Suez to Bombay by steamer in 13 days, Bombay to Calcutta by rail in three days, Calcutta to Hong Kong by steamer in 13 days, Hong Kong to Yokohama by steamer in six days, Yokohama to San Francisco by steamer in 22 days, San Francisco to New York by rail in seven days, and New York to London by steamer and rail in nine days. With setbacks and improvisations, and finding romance and adventure along the way, Fogg managed to win his bet by circumnavigating the globe in 80 days. He gained one day by traveling eastward, as had the clipper ships.

Many miraculous marvels of modern travel made Fogg's journey at least partially credible. Steamships had developed from vessels with side paddle wheels augmenting the sails, to ships driven entirely by steam. The steam pushed giant cylinders and turned underwater screws to drive the ships through the water. Now that the Suez was open and operating, steamships could make point to point runs cutting weeks from the sail routes around the Cape of Good Hope. The use of compound cylinders, where the steam that had driven one cylinder passed on to drive another, had cut the use of coal by up to 50%. This provided more cargo space and the big sluggish ships plodded profitably to and fro between the world's ports.

Another development in technology had been moving forward during the nineteenth century. In 1804, the steam locomotive had been invented and railroads were being built in the United States and England by 1831. In the early 1860s, the United States government proposed developing a railroad across the country. When this thin ribbon of steel was completed on May 10, 1869, a big part of the world was linked together by the speeding steam engines. Another major cut in distance around the world was made in 1855, when a railroad across the Isthmus of Panama was completed.

By 1870, most of the European rail system had been built, and scheduled train service linked major population centers in the industrialized nations. Yet, even Fogg had to resort to travel by lumbering elephant in the underdeveloped world of Asia of the 1870's. Jules Verne's 1873 novel had unknowingly set the stage for the fall of the 35 year record of the James Baines.

The next record holder would not be a tough, hard driving captain of a fast clipper ship, but a diminutive and attractive young woman, named Elizabeth Cochrane. She was born in 1867, in Cochran's Mills, Pennsylvania. At the age of 18, she was angered by the chauvinistic editorial stance of the Pittsburgh Dispatch. She wrote a stinging letter to the editor. The editor liked her writing and agreed to hire her. She asked him for the opportunity to write a series on a taboo subject spoken of in hushed tones, divorce. Elizabeth had chosen her topic well. She adopted the pen name of Nellie Bly from an old Stephen Foster song and proceeded to write a deep and sensitive series on the tragedy of divorce. Papers sold in Pittsburg, and Nellie Bly was soon earning five dollars a week. The city buzzed with speculation of whether the name represented a woman or was the pseudonym for a brilliant male reporter.

Nellie continued to demand and get tougher assignments. With a crusading reporter's zeal, she exposed slum conditions in Pittsburg and sparked reform. She travelled to Mexico and uncovered political problems that had been hidden from the American press.

Before long, she knew she had outgrown Pittsburg, so she packed her Mexican notes, and headed for New York. She needed another bold plan if she was to find a job at Joseph Pulitzer's bastion of male exclusivity, the New York World. With the same brashness she had used so well in

Pittsburg, she tackled the World, marching in on Pulitzer and his Managing Editor, John Cockerill. Pulitzer was impressed by her Mexican work, but got right to the point. What subject could Nellie Bly write about that the World would be interested in?

Nellie had a plan.

"I want to feign insanity," she said.

According to her plan, she could get herself committed to the asylum on Blackwell's Island, and write first hand on conditions there. It worked. She had a job.

Carefully, she removed her identification, and checked into a poor rooming house. She successfully pretended insanity and was committed by a judge to the Blackwell Island Asylum. Rival newspapers played the story to the hilt, speculating on the identity of the poor, unidentified and insane young woman, concocting romantic stories to explain where she came from. The World remained silent on the matter.

For ten days, Nellie lived the horrors of the hopeless people in the asylum, where indifference and cruelty characterized the doctors and nurses. Then, on Saturday, October 8, 1887, the World broke their secret. In the upper left hand box on the front page, the paper declared:

A Strange Phase of City Life:

THE MYSTERIOUS GIRL WHO PUZZLED THE DOCTORS AT BELEVUE/SEE THE SUNDAY WORLD.

Suddenly the mystery spawned by the rival papers was capitalized on by the World and a scathing series of stories followed. Nellie Bly's New York career had begun, and more exposes of the city's sweatshops, and Tammany Hall politics were to come. Then, in October of 1889, Nellie happened on to a new idea. One of her favorite books had been *Around the World in Eighty Days*, and she took the idea from the book to her editor. It was not until a month later that she was told to prepare immediately for a new record breaking trip around the world.

Her preparations bore little resemblance to the lengthy, involved provisioning that Magellan undertook over three and a half centuries earlier. She went to the fashionable dressmaking establishment of William Shormley, and ordered a blue broadcloth dress capable of holding up under constant wear, and a satchel, which she packed with slippers, toilet articles, paper, ink, and other necessities. She took a flask and drinking cup, a 24 hour watch set to New York time, a passport, 200 pounds in English gold and Bank of England notes, and some American money to test its recognition in faraway lands. She carried her valuables in a bag around her neck.

Nellie declined to carry a pistol for her protection, certain of finding gentlemen wherever she might go. On November 14, 1889, Nellie Bly took a horse drawn carriage to the Christopher Street Ferry and crossed the Hudson River. At 9:40 o'clock AM, she set sail on the Hamburg steamer Augusta Victoria for a winter crossing of the Atlantic. While her mother and well-wishers waved good bye, the Augusta Victoria cast off and headed down the bay.

Once at sea, the ship began to toss about in the waves. Nellie started her triumphant journey by becoming seasick. One male passenger laughed, "And she's going around the world alone!" Nellie crawled off to bed in misery. She slept from 7:00 PM on the 14th to 4:00 PM on the 15th. The seas became so rough that two sailors were washed overboard. The World played the event for all it was worth, while preparing the readers for the possibility of failure. There could be absolutely no certainty that the schedules were correct, or that the trains and steamships would leave on time. There was also danger of shipwreck, fevers, and other hazards of travel.

As a professional observer of her surroundings, Nellie noted the man who took his pulse after every meal, the man who counted his steps as he paced the deck, and the woman who slept in her clothes in order to be respectfully dressed if the ship went down. She adapted quickly to the rough seas. This was one of the roughest crossings in years. After six days and 21 hours, the Augusta Victoria docked at Southampton. At 2:30 in the morning, Nellie boarded a tug boat for shore. She was met by the World's London correspondent, who handed her a cablegram from Mr. and Mrs. Jules Verne inviting her to visit them in France. She decided to go without sleep to attempt the side trip.

All of the regular London trains had left, but fortunately a passenger coach had been added to a special mail train. Nellie sent a cablegram back to the World and she and the World's London correspondent were locked into the cold, slow moving passenger coach with a few other passengers. At dawn, they reached the nearly empty Waterloo Station. In a rented carriage, she rushed to the Consulate to pick up her passport and credentials, before she and her companion had a breakfast of ham, eggs, and coffee. Nellie found London streets to be clean compared to New York. When they arrived at Charing Cross Station, the train for Folkstone and Boulonge was pulling out. With a scream, Nellie raced for the rear coach.

Her last minute sprint had saved the day, and Nellie was able to cross the English Chanel in a decrepit boat and catch the train for Amiens. After two days without sleep, the disheveled Nellie Bly got off the train to be greeted by the Jules Verne and a Paris journalist.

From the station, they went to the extensive Verne estate. The bone weary woman did not miss enjoying her visit with the creator of Phineas Fogg. She related that she found both he and Mrs. Verne charming.

Although age and poor health had slowed him down, Jules Verne had travelled much in his life. Pleasantly, they speculated on how Nellie might find a handsome husband on her journey as the fictional Fogg had found a young widow in India. The interview had to be cut short out of deference to critical train schedules. Verne ended with a toast. In English he said, "Good luck, Nellie Bly." Mrs. Verne kissed her on both cheeks and she set out in a carriage for the railway station at Calais.

The Calais train to Brindisi, Italy was the short and fast way across the European Alps. As the train rumbled across Italy with the landscape covered by dense fog, fan mail poured into the World headquarters in New York. Songs were composed, games invented, and promotions resembling today's well-orchestrated hypes were underway. The World suggested she be called Nellie Fly, not Nellie Bly. Meanwhile, Nellie was able to get some sleep in the berth, but it was necessary to pile all of her clothing over her because of the freezing cold. When she stopped at a station and told an Italian cook she was American, he was shocked. He had thought only Indians lived in America.

The train pulled into Brindisi only two hours late. Each stop was a crisis for at each stop a critical connection must be made. She must pick up a ship to the Orient at this point. Fortunately, one was waiting for passengers and ready to sail. It was the Pacific and Oriental steamer, the

Victoria. Nellie cabled New York, after first explaining to the incredulous operator just exactly where New York was.

She began her next leg of the journey in an English ship, the namesake of Magellan's surviving flagship.

The Victoria's crew was arrogant and filled with British chauvinistic pride which rankled Nellie Bly. The food and coffee were cold and indifferently served, and the cabins were poorly planned, lacking light and air. Some of the quarters grew so hot in the Mediterranean that passengers had to sleep on deck. The sailors wore turbans, were barefoot, and sang strange, exotic songs.

Nellie's roommate was a pretty New Zealand girl and they enjoyed each other's company during the voyage. On the afternoon of November 27, 13 days after leaving New York, the Victoria anchored at Port Said. Half naked Arab boys in small boats swarmed around the ship begging for pence. Here, in impoverished Egypt, Nellie saw the great ditch that made her record possible. She was told that building it had cost the lives of a hundred thousand laborers. The canal was a single ditch cut through high sandbanks. Ships were limited to five knots per hour so their wake would not wash out the sandbanks. The steamer Victoria took 24 hours to pass through the canal, a passage that cut weeks and thousands of miles from earlier routes around the world. The huge land mass, of Africa no longer stood between the seaways from west to east.

Nellie was merely a passenger and could only sit back as the Victoria plodded slowly through the Red Sea in the heat. At Aden, she left the ship briefly. Clearly, it was a British world in 1889. British flags and soldiers seemed to be everywhere and Queen Victoria's picture dominated many scenes. English money was accepted. American money was rejected. Nellie had not seen an American flag since she left home. "God Save the Queen" was sung nightly aboard the Victoria, which carried her on to Colombo, Ceylon.

Ceylon was another port, another connection, another crisis. The China bound steamer Oriental could not sail until another steamer, the Nepal arrived. No one could say how long the delay would be.

Nellie enjoyed her stay in Ceylon. She stayed at the Grand Oriental Hotel and enjoyed good food and an after dinner drive or bicycle ride. An Englishman who ran the local newspaper showed her the sights.

Colombo was a picturesque place, with smooth streets laid by convicts and beautiful temples and botanical gardens.

Pleasant as her stay was, the delay stretched out and threatened to destroy her project. By the time the Nepal arrived, Nellie had been stranded five days in Ceylon.

The steamer Oriental was smaller than the Victoria, but Nellie liked it far better. The cabins were well designed and the crew was much more polite. With good weather the ship moved quickly to Penang on the Malay Peninsula, and after a six hour stay, cruised on to Singapore in just two more days. At Singapore, she went ashore for a dinner at the Hotel de L Europe, and a rickshaw tour. The terrible filth and squalor of Singapore was the widest cultural gulf of her journey, as she compared it to the relatively humane world where she had so often been an advocate for improved conditions. She purchased a small Singapore monkey, which she kept as a pet for the rest of her trip.

The Oriental sailed on for Hong Kong at night. Once at sea, a monsoon struck the area. The ship struggled against heavy rolling seas and fierce headwinds. Foaming seas washed over the decks, and water filled Nellie's cabin. In spite of the bad weather, the Oriental arrived in Hong Kong two days ahead of schedule. On this leg, the Oriental not only made up the five days lost in Ceylon, but gained two days. She had broken all previous records for the trip, and Nellie had been lucky enough to be aboard on her mission for speed.

In faraway New York, the World became concerned. They were aware of the monsoons and bad weather, and they were confused by the reports. The editors had received a cable from Colombo that indicated she had been detained five days. Another cable came from Singapore, and another indicated she was in Hong Kong two days sooner than predicted. By Christmas, no one knew where Nellie Bly really was, and it seemed impossible that she could be where her cables were reporting.

Nellie had reached the point of a new connection and new problems. A steamship office agent informed her that she would have a five day layover here and a five day layover in Yokohama. He also told her with sadistic glee that the World had sent another woman to race against her, and that the other woman was already ahead. Crushed by the thought, the level headed young woman came near panic. Fortunately, the purser from the Oceanic found her near this time, after searching out her whereabouts. The Oceanic was the steamship which was to carry her to Japan and the United States, and its plans were not nearly as dire as the agent had prophesied.

Her Christmas present was to see the United States flag again, flying above the Consulate in Canton, the first time she had seen the stars and stripes since leaving New York.

Hong Kong and Canton had offered nothing but sights of misery and poverty. Her brief tour of China was one of grim enlightenment. Canton had over 800 temples, including the Temple of Horrors. This was filled with carvings of people undergoing sundry tortures and executions. It was surrounded by lepers, cripples, gamblers, and fortune tellers. She had seen leper villages and prisons where the prisoners were systematically tortured. In one court, Nellie watched judges stop smoking opium long enough to pronounce sentence on two prisoners. They were chained with their knees under their chins and carried in baskets by coolies. The sentence was to have every bone in their hands crushed by heavy stones.

She was shown the execution grounds with many crosses. The guide explained that condemned women were tied to the crosses and cut to pieces. Men were quickly beheaded except for the worst criminals, who would suffer the death of a woman. This commentary from the guide did little for Nellie's sense of equality. She had to return to Hong Kong, as foreigners were not allowed to stay in Canton overnight. Although she had a fairly pleasant Christmas at the American Consulate, she was more than happy to leave for Japan. After transferring baggage and monkey from the Oriental to the Oceanic, she set out into the Pacific for Japan on December 28, 1889.

New Year's, 1890, was welcomed in at sea aboard the pitching Oceanic. On January 3, Nellie Bly arrived at Yokohama. Japan was a contrast to the lugubrious society in China. Life was upbeat and simple, a blend of East and West.

"Japan is a load of beauty, love, poetry, and cleanliness," Nellie wrote. "The Japanese are a cheerful, happy people. Their food is excellent, and their dainty houses clean and doll like."

A Japanese newsman showed her a copy of the New York World with her Jules Verne interview. She began to realize the stir her trip was creating in the states and even in Japan.

The Oceanic was not scheduled to leave until January 7. The mails it carried were not ready and there was no way it could leave sooner. Nellie had four days to tour Tokyo and Yokohama. When the Oceanic left the harbor on the 7th, the Chief Engineer had written a message and placed it over the engine:

For Nellie Bly We'll Win or Die

The steamship made good time until the third day out. The vast ocean that had nearly destroyed Magellan's expedition now fought back at the comparative ease and speed with which man crossed her expanse. The Oceanic found herself struggling against powerful headwinds and high seas. Mother Nature had unleashed a fierce winter on the world in 1890. Blizzards stranded trains in deep drifts in the western states of the United States. Cattle died in droves on land, while the Oceanic bucked and tossed in the Pacific, fighting against the January storm with all of the modern technology built into her. She survived and pulled into San Francisco harbor to meet another connection and another immediate difficulty.

Someone had started the rumor that there was smallpox on board the Oceanic. No one could leave until the ship's purser could produce a bill of health. Unfortunately, he had left it behind in Yokohama and it would be two weeks before the next boat from Japan would arrive.

"I'll jump overboard and swim ashore!" Nellie proclaimed, and her bearing told her listeners she meant it. Somehow, the paperwork was sidestepped and Nellie was allowed to board a tugboat for shore with her bag and her monkey.

Large cheering crowds had gathered, including the mayor and the Press Club. After a short speech, Nellie was rushed to the Atlantic and Pacific Railroad station to board a special train for the final run home.

Travel across the United States in the winter of 1890 by rail was not completely without problems. Snow blind or dead cattle blocked tracks, and some rerouting had to be done to avoid blizzards. Telegraph lines and many of the less prevalent telephone lines had been disrupted by the storms. At every small town and station people gathered to wave and cheer as the train rumbled by. At one point, the train nearly derailed when it crashed into an empty sidecar left on the tracks. At another point, near Gallup, New Mexico, the train surprised a group of track repairmen working on a bridge over a 100 foot ravine. Unprepared, and without sufficient time to give warning, the crew could only watch as the train roared across the bridge which was supported only by jackscrews. The bridge bent under the weight of the heavy locomotive, but it held.

The train rumbled and chugged through New Mexico, Arizona, and Colorado, at the fantastic speed of 60 miles per hour. What would the early circumnavigators have thought of that? Indians waved from their ponies. Ranchers came for miles with their families to watch her train pass by. In Kansas, suffragettes begged her to return and run for governor.

In Chicago, the Press Club provided Nellie with a breakfast reception between trains. Amid the ceremony, she was handed a message from Jules Verne wishing her many good things. The book, *Around the World in Eighty Days*, had gone into its tenth printing, and had been revived as a play in Paris. If Nellie could beat Phineas Fogg, she was to be written into the epilog of the play.

At Pittsburgh, she was able to wave to many of her friends, from the old newspaper crowd other hometown and at Philadelphia, her mother and several friends boarded for the final leg of the journey. At Jersey City, her mad rush around the world was complete. As she stepped off onto the platform, three stopwatches recorded the moment as 3:51 o'clock PM.

From Jersey City to Jersey City, she had gone around the world, breaking every previous record. She had made it in 72 days, six hours, ten minutes, and eleven seconds. Cannon boomed from the Battery and Fort Green Park in Brooklyn. Amid hysterical crowds and flowery speeches, Nellie came home, a triumph of one woman's determination and nineteenth century technology.

Nellie had not met her true love on her journey, but her future would include marriage, widowhood, running a factory, and more reporting. She died of pneumonia at the age of fifty-four in 1922.

This trip was not like the elaborately prepared and financed expeditions of three centuries before. It was not like the rugged sailing journeys of three decades before. It had cut in half the fastest time for a circumnavigation, and beaten the fictional trip of Phineas Fogg. Nellie had combined steamship, horse carriage, and train and had traveled by purely public transportation, even though special efforts had been made on her behalf. This trip was to spark interest in fast circumnavigations and would result in others following her lead.

This trip was the first record utilizing more than one means of transportation. It was the first record trip to cut through the Suez Canal, to use trains, or to use passenger steamships. It was the first record set by a woman alone, and it was the first to cross the North American continent and the European Continent. Man was no longer bound to the water in order to travel long distances quickly.

"The Amazing Nellie Bly" set a record, but her publicity would encourage fast followers, and in less than a year others would be trying to break her record time. Her trip would remain remarkable for a long time. It would be 46 years before there would be a record breaking circumnavigation that would cover a similar distance.

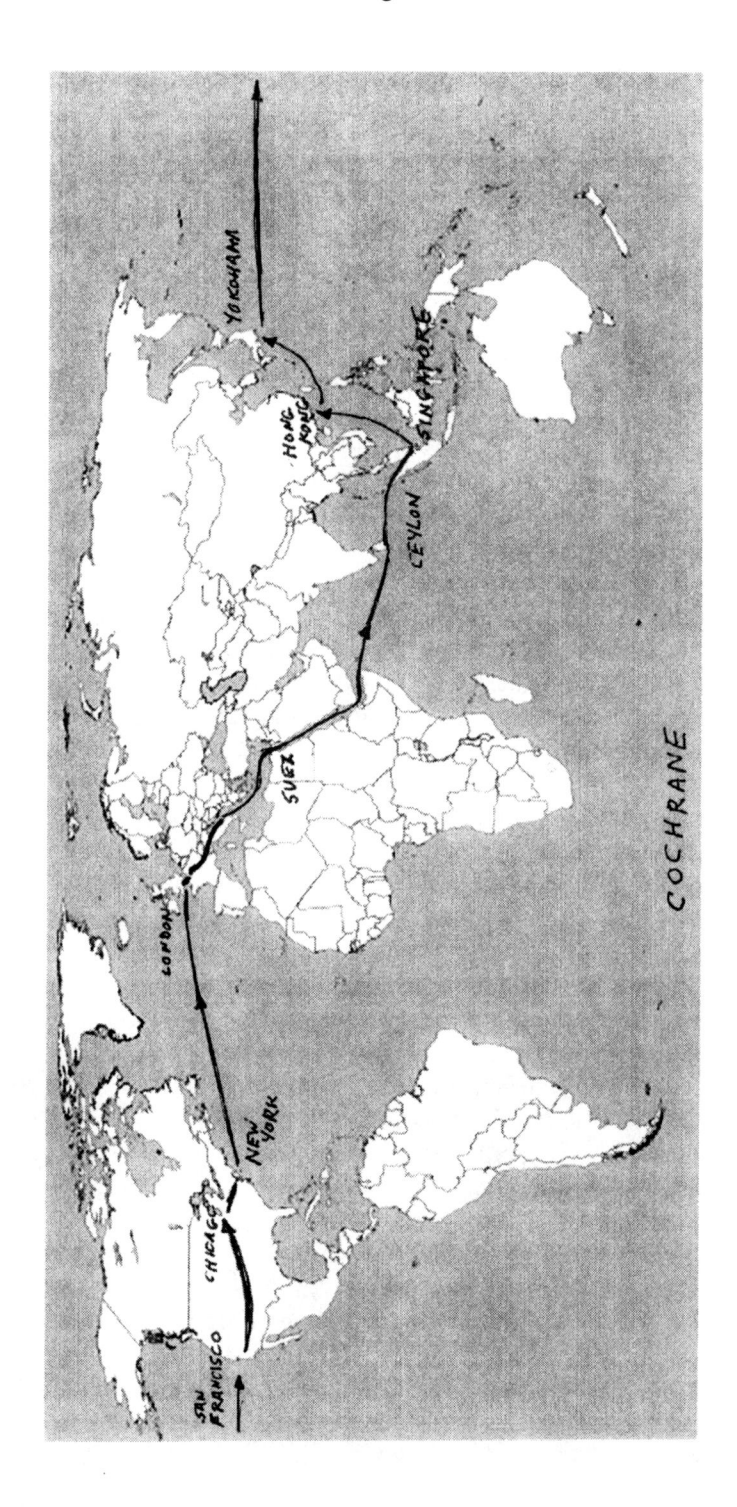

Circumnavigation

Alan Boone

CHAPTER 10 - ALSO RANS

The first globetrotter to follow Nellie Bly's example and try to beat her record was a man named George Francis Train, who traveled the world to promote his hometown of Tacoma, Washington, and ended up cursing the name of his own city.

Train, apparently a man of considerable means, traveled from east to west. With much fanfare, the flamboyant Train left the Tacoma Hotel on March 18, 1890, and was quickly taken to the steamer Abyssinia. Train remarked that he was fixed so that he could buy a steamer if his connections failed. The Abyssinia was not the fastest ship to Yokohama, but he hoped to make up time with a special train from New York to Tacoma. His declared goal was 6o days around the world.

At 61 years old, Train was no stranger to world travel. He had been very much involved in the fast paced clipper ship runs of 35 years earlier by owning his own shipping firm and making a quick fortune during the Australian gold rush.

This time, travelling under the auspices of the Tacoma Ledger newspaper, he reached Yokohama in 16 days, a run that was slower by ten days than a steamer doing normal time. Using characteristic bravado, he succeeded in getting a passport in 20 minutes and holding the North German Lloyd steamer, General Werder, in Kobe by wire. Crossing Japan by special train, he caught the ship, but was delayed two more days getting into Hong Kong by a situation that even a rich man couldn't buy out of, a heavy fog.

Train left Hong Kong on April 13, following a route similar to Nellie Bly's, only in reverse. That route took him to Singapore, Ceylon, Aden, Suez, and across the Mediterranean to Brindisi, at the bottom of Italy's boot. He continued across Europe by train, to Paris. Crossing to London, he caught the steamship Etruria at Queenstown, Ireland, bound for New York.

In New York, Train breakfasted at the Astor House, exciting press comment about his flashy dress and constant plugging of Tacoma.

By the time he returned to Tacoma, however, his tune had changed. Train had traveled from New York to Portland, but had been robbed of $600 and his transportation while in Huntington, Oregon, the day before. Upon arriving in Portland, the expected special train, which was to be provided by the city of Tacoma via the Northern Pacific Railroad, was not available. Train had to take a Union Pacific coach instead. He paced up and down the aisle of the Union Pacific car in his white suit and tropical hat, loudly and profanely denouncing Tacoma, and those associated with the trip. The New York Times reported that Train was "The most disgusted globe trotter in history."

It was 6:45 PM, May 24, when the Union Pacific delivered the raging Train to his home.

Despite his nasty temper, he would have succeeded in breaking Nellie Bly's record of less than a year before, but his distance of 22,140 miles was only a few miles shy of the distance necessary to qualify as a true circumnavigation. His return to Tacoma was only 67 days, 13 hours, three minutes and three seconds, in spite of a 36 hour voluntary layover in New York for breakfast and press play.

Train repeated his global trek the next year, beating his old record and lowering his time to 64 days. This time, he was publicizing New Whatcom, Washington. The eccentric Train offered an around the world trip as a publicity stunt to the newly reorganized Chamber of Commerce of that city.

The city raised a thousand dollars for the trip and sent him off on May 9, 1891, on the steamer Premier to Vancouver. At Vancouver, he caught the Empress of India to Hong Kong, where he missed an important connection by just twenty minutes. By June 4, he had reached Singapore. From there, he passed through the Suez Canal at Port Said and landed at Liverpool just 45 days after departure. Six days later he was in New York. From New York, Train travelled by rail through Chicago and Omaha, to reach New Whatcom.

Upon his return, he was able to present clippings from newspapers in the Orient, New York, London, Omaha, and Chicago, each mentioning New Whatcom. The populace had difficulty seeing that they had received an adequate return on their investment, and his arrival is reported to have been less enthusiastic than his send off.

Eleven years after Train's first circuit, a young man who was later to become the Chicago Police Chief, traveled the globe in a little over 60 days.

Charles Fitzmorris was a 17 year old third year student at Chicago's Lake High School, when he was voted Chicago's Most Popular Boy. At this time in 1901, the Hearst newspapers were sponsoring a race around the world. As part of the honor of Most Popular Boy, Fitzmorris won the right to compete with other schoolboy representatives of the San Francisco Examiner and the New York Journal.

At 5:30 PM, May 20, 1901, Fitzmorris left Chicago. He traveled by rail and steamship, covering 10,390 miles by water, and 10,065 miles by land for a distance of 20,465 miles. This distance, like Train's, was short of the Tropic of Capricorn by a small amount.

At 7:00 AM, July 20, 1901, "the fastest horse in Chicago" drew his buggy up to the offices of the Chicago Evening American. Fitzmorris had completed the circuit in 60 days, 12 hours, 29 minutes, and 42 and 4/5 seconds. As with other also-rans, the names of the other two competitors are not mentioned in history.

J. Willis Sayre had started out as a young man by folding programs for the Seattle Opera House, and later became Seattle's first theatrical advertising manager. He served in Manila during the Spanish American War. In 1901, he became Drama Editor for the Post Intelligencer, and in 1903, he set out to take as long a trip as possible during a relatively short vacation time. He decided to circle the globe, travelling only first class, and using only commercial transportation.

Since no passenger ship was available when he was ready to leave, he procured a booth on the freighter, Hyades. The Hyades sailed at 5:55 o'clock AM, Friday, June 26, and crossed the Pacific on the great circle route past the Aleutians in 19 days.

At Yokohama, Sayre crossed Japan by train in time to catch a show in Nagasaki and sail to Dalny, Manchuria. Here, Sayre became one of the first to travel the Trans-Siberian Railroad in its first week of operation. At a Russian restaurant, the somewhat naive Sayre held up two fingers and cackled.

"Oh, you want two eggs?" the waiter said in perfect English.

The Trans-Siberian stopped three times daily for everyone to get out and cook their meals. Having successfully traversed the vast Russian flatlands to Moscow; Sayre took the wrong train out and had to take a farm wagon back to Moscow to catch the right train to Poland. He was the only first class passenger and had an entire car to himself on the way to Warsaw. At the German border, he had a delay while he struggled with the German bureaucracy, losing a full day of travel time.

A Dutch liner took him to England and he was able to catch the Campania at Queenstown, Ireland for New York where he boarded a train for Minnesota. At St. Paul, he received a cable offering a special train to Seattle, a gift from hometown friends. Sayre refused, however, and took the Northern Pacific home. He had officially been clocked at 54 days, nine hours, and 42 minutes, a record that would stand for a scant week.

Although the Trans-Siberian Railroad shortened the distance and time around the world, it also cut distances short of that necessary to qualify as a true circumnavigation.

Henry Frederick left New York City aboard the steamer Deutschland on July 2, 1903, concurrent with Sayre's crossing of the Pacific in his attempt at a record. The steamer crossed the Atlantic in the quick time of six days. From Paris, Frederick spent 18 days by rail to Dalny, Manchuria. His time consisted of two days to cross the Yellow Sea, two days by rail across Japan, and 16 days on a slow steamer over the Pacific. The American continent was crossed in just four days.

While crossing Asia, he faced the evidence of growing international friction between Russia and Japan, with critical bridges and passes guarded by well-armed detachments of soldiers. In Yokohama, Japan, he missed his connection, resulting in his slow crossing of the Pacific. His only night in motionless sleep had been in a hotel in Yokohama.

When the New York Central Railroad brought Frederick home, he had beaten the time of Sayre by just two hours and forty minutes and established a record of 54 days, seven hours, and two minutes. Both Sayre and Frederick had taken the Trans-Siberian Railroad, and their full distance fell short of true circumnavigation.

On June 18, 1907, Lt. Colonel H. Burnley-Campbell wrote a letter to the editor of the London Times which was published on June 22. He described a journey via the Trans-Siberian Railway in which he covered the distance around the upper part of the world in forty days, 19 1/2 hours. He had succeeded in making good connections and ran into trouble only once, when his ship ran aground during a dense fog in the Sea of Japan. It was later floated off undamaged at high tide.

Burnley-Campbell sailed from Liverpool toward the west aboard the Empress of Ireland on May 3, 1907, at 7:20 o'clock PM, and reached Quebec at 3:00 o'clock PM, on May 10. Two hours later, he started across Canada via the Canadian Pacific Railway, arriving at Vancouver, May 14. He sailed on the Empress of China and reached Yokohama May 26. He crossed Japan by rail to Tsaruga and picked up a Japanese steamer to Vladivostok from May 28 to May 30. He boarded the Trans-Siberian train which passed through Harbin May 31, Irkutsk, June 4, and Moscow, June 10.

Continuing by rail, he passed through Warsaw June 11, Berlin June 12, Cologne June 12, and Ostend June 13. On June 13, at 2:50 PM, he arrived at Dover. His letter of June 18 inquired if he had set a record. Indeed, he had, for the particular distance covered.

The Frenchman Andre Jaeger-Schmidt traveled 19,300 miles in 1911, for the Paris newspaper, Excelsior. His trip took him from Paris to Vladivostok, Japan, Vancouver, Montreal, New York, Cherbourg, and he returned to Paris in 39 days, 19 hours, 43 minutes, and 37 4/5 seconds.

This was followed in 1913, by John Henry Mears, an American, who traveled 21,066 miles in 35 days, 21 hours, 43 minutes, averaging 24 miles per hour. Mears left New York on July 2, taking the Cunard liner Mauretania to England. He traveled via London, Paris, and Berlin to pick up the Trans-Siberian Railway from St. Petersburg to Vladivostok, switching to the South Manchurian and Chosin Railways to travel through Manchuria and Pusan, Korea.

Catching the Empress of Russia from Yokohama to Victoria, British Columbia, the 38 year old Mears switched to rail at Seattle and passed through Chicago and Buffalo to reach New York. The trip, which lasted a little over a month, was barely short of the distance needed to qualify as a record for purposes of this book. It was a significant trip, because it marked the end of record breaking travel by car, train, and steamer. From then on, man would have to take to the air to go faster.

While steamship lines and railroads spanned the globe, men were working with a new contraption which moved through the air. Early aeroplanes were flimsy wood and fabric machines resembling box kites. They could take you into the air but do little more.

World War I hastened the development of aircraft technology, and soon aircraft were attaining greater speeds, altitudes, and payload capacity. They could drop bombs on armies from the skies, and engage in high speed machine gun fights in the clouds.

Although development slowed after the war, airplanes with big, powerful engines, comfortable cabins, and moderate overland ranges were soon carrying paying passengers from city to city.

The first flight around the world was accomplished by U. S. Army planes in 1924, but the time could have been beaten by the fast sailing ships of 50 years earlier. They covered 26,103 miles and 21 countries in 57 "hops." This trip took six months and 22 days.

Using planes, trains, and autos, Edward S. Evans and Linton Wells of the New York Globe circled the world in 28 days, 14 hours, 36 minutes, and three seconds, but traveled only 18,400 miles from June 16 to July 14, 1926.

Then, in 1928, John Henry Mears once again returned to globetrotting. Two great barriers, the Atlantic and the Pacific, stood between the circumnavigators and aircraft speed. Flying machines of the day lacked sufficient range to cross oceans. This problem was solved by Mears and his pilot, Charles B.D. Collyer, by shipping their Fairchild cabin monoplane across the Atlantic on the steamer Olympic, and across the Pacific aboard the Empress of Russia. Travelling from west to east, they landed at Cherbourg, where they off loaded their airplane and flew over Europe and Asia to Tokyo in just six days. Crossing from Tokyo, Japan, to Victoria, Canada, on a steamer, they again took to the air for New York. After 15 days, 8,535 miles on ships, and eight days, 11,190 miles in flight, they had circled the northern part of the world, covering 19,725 miles in 28 days, 14 hours, 36 minutes, and four seconds.

As the airplane was developing during World War I, lighter than air craft were being successfully used by Germany. Early craft consisted of balloons filled with a gas which was lighter than the air which surrounded it, causing it to float above the ground.

Soon a rigid structure was built to give the balloon a distinct cigar shape. The rigid shape gave this kind of airship the name dirigible. Although experiments with airships had been carried out as early as 1852, it was the German Count Ferdinand von Zeppelin who made the airship practical. Prior to the war he had successfully operated a commercial airline with these craft. For many years dirigibles were called Zeppelins after the innovative Count. They would remain in service until 1937.

The dream of travelling the world totally by air was realized when the Graf Zeppelin left Lakehurst, New Jersey, on August 8, 1929. She traveled to Friedrichshafen, Tokyo, Los Angeles, and returned to Lakehurst. She had circled the world in four giant hops in 21 days, five hours, and 54 minutes. After a short rest, she repeated her feat, this time lopping one day, one hour, and forty minutes from her time. The Graf Zeppelin covered 21,700 miles, just short of covering the necessary distance.

In 1931, Wiley Post and Harold Getty flew the Winnie Mae over the Arctic Circle route. They flew New York to New York from June 23, to July 1, 1931, via Newfoundland, Chester, England, Hanover, and Berlin, Germany, Moscow, and several other Russian cities, Fairbanks, Edmonton, and Cleveland. They covered only 15,474 miles in eight days, 51 hours, and 51 minutes. Wiley Post became the first to fly the Arctic Circle solo two years later. He covered 15,596 miles on that trip.

These had been special trips with special equipment. In 1936, H. R.

Ekins was to test the state of regular air travel around the world, and he would be the first person since Nellie Bly to set a record for a full trip of over 22,859 miles.

Alan Boone

CHAPTER 11 - EKINS

On the dark rainy evening of September 30, 1936, the Hindenburg sat in the floodlights at its dock in Lakehurst, New Jersey. The imposing 812 foot long zeppelin, one of the largest ever built, contained over seven million cubic feet and was filled with highly flammable hydrogen. The airship was destined to become a victim of tense relations between the war posturing Germany and the cautious United States. We had the less flammable and more stable helium: but we refused to sell it to the German government, or their private enterprise. This forced them into the use of the unstable hydrogen to fill their giant airship. The rigid zeppelin was soon to cross the Atlantic Ocean. The two great expanses of water that had carried the early circumnavigators had stood in the way of around the world travel by air. Now, they were bridged by zeppelins and big four motor clipper flying boats. These large aircraft took off and landed in water and had hulls like ships. In 1936, just prior to the world shattering war, it had become possible to traverse the entire globe by scheduled air service. No longer was it necessary, as it had been just eight years earlier, to transport an airplane across the sea by steamer to complete a fast circumnavigation.

The recently developed ability to travel quickly and comfortably by air sparked the interest of a Scripps-Howard Newspapers publisher. He assigned H. R. Ekins to undertake an air trip and set a ground rule of using only established, scheduled airlines.

After securing more than a dozen visas and a batch of airline tickets,

Ekins met his first disappointment. He had planned to take the American Airlines shuttle aircraft on a twenty minute flight from New York to Lakehurst. A zero ceiling at Lakehurst however, forced him to use a three car motorcade. He left New York at 8:17 PM, and arrived in Lakehurst a depressing two hours later. Air travel, he could ruefully note, was modern, but no match for an uncooperative Mother Nature.

Two other newspapers had entered into an impromptu race with the World Telegram. The New York Times was sending Leo Kieran and the New York Journal was sending one of its young women reporters, Dorothy Kilgallen. The three were among the forty passengers on board the Hindenburg when she sailed on September 30, 1936. The term sailing was used to refer to the launching of these airships, just as it was used for steamships. It was a carryover from the days of sailing.

Ekins had prepared well. His arrangements included several alternate bookings in case he arrived late at any location. Armed with his stack of tickets, timetables, a small amount of luggage, and three wristwatches, one of which was set for New York time, he rode an airport bus to the Hindenburg, and climbed the swaying gangway into the dirigible's belly.

The Hindenburg was late in launching that evening. She got underway at 11:17 PM, with the release of a giant stream of water ballast. Slowly, quietly, she rose straight up into the rainy skies. There was no noise, no vibration, and no sway. The engines pushed her out toward the Atlantic at an altitude of 600 feet, which she maintained throughout the trip. The race participants gathered in the asbestos smoke room and bar to discuss arrangements and schedules for the rest of the trip, and then turned in for the night.

The next morning, the dirigible was flying under gray skies and heavy rain squalls. At times, she slowed to 30 knots, but by afternoon, she had speeded up to 46 knots. While the seas roiled and boiled beneath her, the passengers floated in comfort with soft lounge chairs, thick carpets, fine cuisine, and aged wines. At night they slept in comfortable staterooms.

By the evening of the second day, the airship was making 68 knots, and was only 500 miles from the English Channel. It was becoming apparent that Ekins was going to miss his first vital connection with a Lufthansa aircraft. The poor weather early in the flight had stymied progress. Fortunately, his backup reservations aboard a KLM flight would solve his problem. The Hindenburg cruised across the checkerboard landscape of Europe to Frankfort and moored at her home base, opening the hatch to the gangway at 3:19 PM.

After a quick glass of German beer, Ekins rushed across the field to the KLM plane, a 14 passenger Douglas DC-2. Elkins was on the ground exactly one hour before again becoming airborne, leaving his competitors far behind. Ekins had a distinct preference for the American built DC-2, with its two 890 horsepower Wright Cyclone engines. This plane, named the Flamingo, flew out over medieval Nuremberg at 150 miles per hour, then flew over the snow covered forests of the Danube valley. Ekins witnessed his third sunset since leaving New York. The Flamingo landed at Vienna at 6:41 PM, after two hours and 28 minutes of flight, touching down at Aspern Airport 396 miles from Frankfort.

Ekins spent a pleasant and late evening in Vienna with some long time newspaper friends and became convinced that the greatest hazard facing the world traveler in modern day 1936 was the hospitality of friends and acquaintances. They made short work of long hours and made leaving a bittersweet task. This was his first night on the ground and it was a short one, for at 5:14 the next morning, the Flamingo took off for Athens.

Playing tag with the Danube, the plane passed above the clouds over Hungary and Yugoslavia. It bounced over the rough air in the mountain valleys, clearing the top of a mountain pass by only 500 feet at one point. The airplane provided a speedier and rougher ride than the dirigible. Only the great Pan American clipper aircraft in the Pacific could match the dirigible's range, however.

When the DC-2 landed at Athens, she had covered 843 miles non-stop in five hours, 45 minutes. Ekins had to spend another night on the ground here. He spent it sampling the Athenian night life with the fourth estate, but he took note of the lesson learned in Vienna, and turned in for a few hours' sleep before his departure for Alexandria, Egypt, at 5:55 AM, Monday, October 5.

Again, he was on a KLM DC-2. This one was called the Kwak or Night Heron. The DC-2 was the forerunner of the famed DC-3. That aircraft, only slightly changed from the DC-2, would be the workhorse of the world airline fleets for many years. The Kwak was destined to traverse 1,800 miles this day, an impressive feat of modern transportation. The speed conscious world traveler noted in his journalist's mind how impressive transportation had become.

Less than four and one half days from New York, Ekins crossed the Mediterranean. The crossing did not take up a full hour, and Ekins soon entered the hot desert world of Egypt. Swarthy men in fezzes and the ubiquitous English soldiers attested to the fact that the British Empire was still very much in evidence, as it had been 46 years earlier when Nellie Bly made the lengthy earthbound trip for her newspaper.

Ekins was on the ground at Alexandria only 34 minutes, before taking off for Gaza. He passed over the thin ribbon of the Suez Canal laid across the sandy wastes of the Sahara. Again, at Gaza, they were on the ground for a brief half hour, then ascended to cover the barren Middle Eastern landscape from the more pleasant perspective of the air. With Baghdad as their destination, they crossed the Dead Sea and Jordan River. The aircraft droned on, stopping temporarily in Baghdad, and then going over the endless looking desert with its monotonous landscape, broken only by occasional oasis and camel caravans. After crossing the Tigris and Euphrates, the airplane sat down at Basrah, Sinbad's port of departure. After taking just 12 hours and 20 minutes to cross from Europe, over Africa, and into Asia, Elkins finally got a good night's rest at Basrah.

Again about 5:00 AM, he was in the air in the same DC-2, crossing the swampy areas to the Persian Gulf. During the four hour flight to Djask, the KLM crew served a "second breakfast."

Djask was little more than a few huts and served as a cable station and refueling stop in the extensive KLM routes. Only a brief stop was made here, then the plane continued on a busy schedule to Karachi. It was 2:55

PM, when the airliner reached Karachi, India. Ekins was briefly in the exotic city, becoming airborne again at 5:15 PM, for the short hop to Jodhpur. Elkins passed over the top of Hyderabad, considered at the time to be the hottest place on earth, with 135° in the shade during the summer. From the rooftops, he could see the tall, protruding ventilators which were designed to catch the slightest breath of air.

At Jodhpur, Ekins again stayed overnight. Lack of beacons made night flying in the East hazardous and resulted in the airlines preferring stopovers in the evenings. This provided Ekins with fleeting opportunities for sightseeing in this town of ox carts, camels, temples, and sacred cows. Six days from Lakehurst, he slept in the heart of India.

His rest was interrupted early, with a pre 5:00 AM departure to the east. Once again, the KLM plane, Night Heron, headed for the skies, leveling off at 5,000 feet. Crossing wild and rocky country, including the Aravilli Mountains, they finally came to an area of fertile tablelands with water and rice paddies and a broad swollen river, the Sacred Ganges.

Touching at Allahabad, the aircraft sped on across the Bengalese plain to the Calcutta Airport with the unlikely name of Dum Dum. Ekins was told that the name came from a munitions plant in the area whose production included the Dum Dum bullets.

The Calcutta stop was only thirty minutes, and the small, sturdy plane was off to cover the 670 miles to Rangoon, on the other side of the Bay of Bengal. At Rangoon, Burma, the Rangoon Times and the KLM officials greeted the traveler with a bevy of dancing girls. Ekins would have another night's rest on the ground in this exotic city.

After the rest in Rangoon, the durable airplane made its customary 5:15 AM departure to the southeast down the rugged peninsula of Sumatra to Singapore.

At Singapore, Ekins spent the night at the Sea View Hotel, typing out his latest dispatches to his paper. He was also able to learn that his rivals, Dorothy Kilgallen and Mr. Kieran, were at Sharjab, Arabia, some 4,000 miles behind.

At 5:35 AM, Captain Rondong, who had flown them 9,000 miles from Europe, headed the Douglas aircraft eastward.

During the next 24 hours, Ekins would breakfast in Singapore, lunch in Batavia, and take his evening meal in Balikh-Papan on the coast of Borneo. The plane roared above fabulous equatorial islands; the spice growing islands that had lured the first circumnavigators. Batavia, the rich capital of the Dutch empire of the Indies, was still a teeming Dutch port, filled with pride at the speed and dependability of the Dutch airlines that had carried Ekins so far.

In Batavia, Ekins changed planes and airlines. Now, he would fly the Douglas aircraft of KNILM or Royal Netherlands Indies Airlines. This plane was called the Pkfal. Ekins was impressed by the civility and modern conditions of Balikh-Papan. A pleasant night's sleep lasted a little longer than usual, since this time the KNILM plane didn't depart until 7:30 AM. Ekins was now headed north toward the Philippines.

Just after leaving the coast of Borneo, the Douglas hit the first serious weather of the trip, bouncing through a blinding rainstorm like a small ship on rocky ocean waves. By late afternoon, Captain Van Bremer brought the aircraft and its weary passengers into Zamboanga. The gales tore coconuts from the trees and bent the hemp fields over with their continuing fury. Van Bremer set his DC-2 with its 90 foot wingspan into a field just 110 feet wide. The field was lined with palm trees and the cross wind was fierce, giving credibility to the skill of flying with the aviators of the day, and pointing to later laws requiring clear zones around air fields and ports.

Now, Ekins' speedy trip was beginning to have obstacles. The storm had delayed the boat carrying aviation fuel to the air field. Scrounging 87 octane gasoline from the Army Air Corps, the DC-2 was refueled with painful slowness from five gallon tins. Through black and stormy skies, the refueled plane bounced its way to the soggy air field at Manila. Ekins had come half way around the world in nine and one half days.

He ceremoniously exchanged letters from President Franklin D. Roosevelt and Mayor LaGuardia of New York City with President Don Manuel Quezon. Now, weather became a dangerous block. A tremendous typhoon lashed the Pacific. For three days, Sunday, Monday, and Tuesday, Ekins remained grounded in the Philippines while his rivals closed the earlier gap. There was finally a sufficient break in the weather to get away. He had sampled the ever present hospitality of the island's people, but it was with relief that he boarded the plane to continue his race.

From the Philippines, Ekins boarded an American airline, Pan American. The plane was vastly different from the cramped passenger cabin of the DC-2. Unlike the other airplanes on the trip thus far, this plane was a flying boat that could land and take off in water. Its airports were ports in the true sense of being harbors. It weighed 51,000 pounds loaded, and its 130 foot length and 92 foot wingspan were carried along by four 1,060 horsepower supercharged Pratt & Whitney engines. She had a lounge that seated 16 people, three other compartments capable of seating 12 or sleeping six, a full galley, a baggage room, and an elevated bridge for the four man flight crew. The craft was named the Hawaii Clipper, and could fly at 175 miles per hour and remain aloft for several hours. Still, the vast Pacific Ocean, the nemesis of so many early circumnavigators, could not be crossed in a leap. The clipper roared for hour after hour, bouncing onto and off of the waves at Guam, Wake Island, Midway, and Honolulu, calm paradise islands that were so shortly to become embroiled in violent, bloody war.

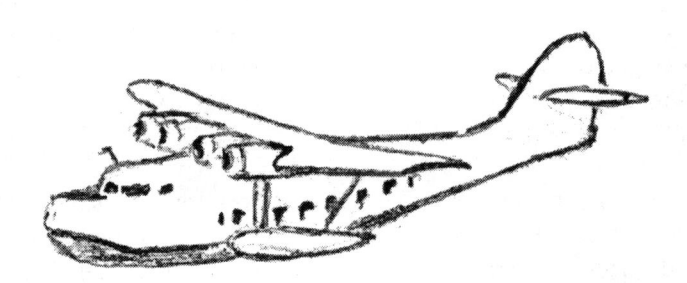

Not being limited to wingspans less than the planting distance of the local airfield flora and fauna, the flying boat quickly refueled near these islands, and then returned to the air like a great white pelican full with its catch.

When Ekins left Wake Island, it was Friday, October 16. When he landed at Midway, it was Thursday, October 15, having crossed the International Dateline. The following morning it was Friday, October 16 again, so on the 16th, Ekins flew from Wake to Midway, and on the second 16th, he flew from Midway to Honolulu.

Leaving Honolulu, the clipper was to make its longest leap, one unparalleled in scheduled flying history up to that time. At noon on Saturday, Captain LaPorte took off for San Francisco, 2,400 miles away. The plane was to fly all night while Ekins slept in the comfortable sleeping berth. By morning, she roared over the newly built Golden Gate bridge. At 10:23 AM, the ground crew in a rowboat drew the clipper alongside the landing dock. Ekins now faced a lengthy bout with customs, then drove to Oakland where he changed to his fifth airline, United.

The new plane was a comfortable Boeing transport for the short distance to Los Angeles. This short leg of the journey took its toll in time. While struggling over the mountain ranges of Southern California in a bad storm, the pilot received word of a lowering ceiling around the hill lined Burbank Airport. He turned back to Bakersfield and waited out the storm. As soon as the weather improved he went on to Burbank, but the trip took more time than some of the hops from country to country.

There, Ekins changed planes and airlines again. He continued his journey east in a Transcontinental and Western Airlines (TWA) DC-2. The familiar craft accommodated smoking and carried a hostess. Ekins was carried across the United States by TWA, stopping at Albuquerque, Wichita, Kansas City, and Pittsburgh, where Nellie Bly had spent her early years. After leaving Pittsburgh, the plane flew for an hour and passed over Lakehurst to log a record of 18 days, 11 hours, and 45 minutes, even though they didn't land. Ekins flew on to Newark Airport and landed at 10:49 AM, five minutes ahead of schedule.

Through a crowd of reporters and greeters, Ekins made his way to the waiting motorcade. With police escort, they rushed over the Pulaski Skyway, through the Holland Tunnel, and across Church and Barclay Streets to the World Telegram offices, where stop watches clicked at 11:14 AM on Monday, October 19. Elkins had left the lobby of the Telegraph on September 30, at 8:17 PM. In 18 days, 14 hours, 56 minutes, and 50 2/5 seconds, Ekins had traveled 25,794 miles around the world, entirely by scheduled air routes. He beat his competitors by six days.

One dirigible and seven aircraft from five airlines had made the distance, passing over land and sea at speeds that would have amazed a Magellan, Drake, or even bold "Bully" Forbes. There had been no scurvy, or lack of food and water. Nor had they been in danger from pirates and natives. The entire trip had been made in the relative comfort of passenger travel. Ekins missed the sights of penguins and Patagonians, and his profit was counted in newspaper subscriptions instead of gold booty or delicious spices, but the adventuresome spirit was the same force that had driven the irrepressible Magellan or vengeful Drake in their quest of bounty from our planet.

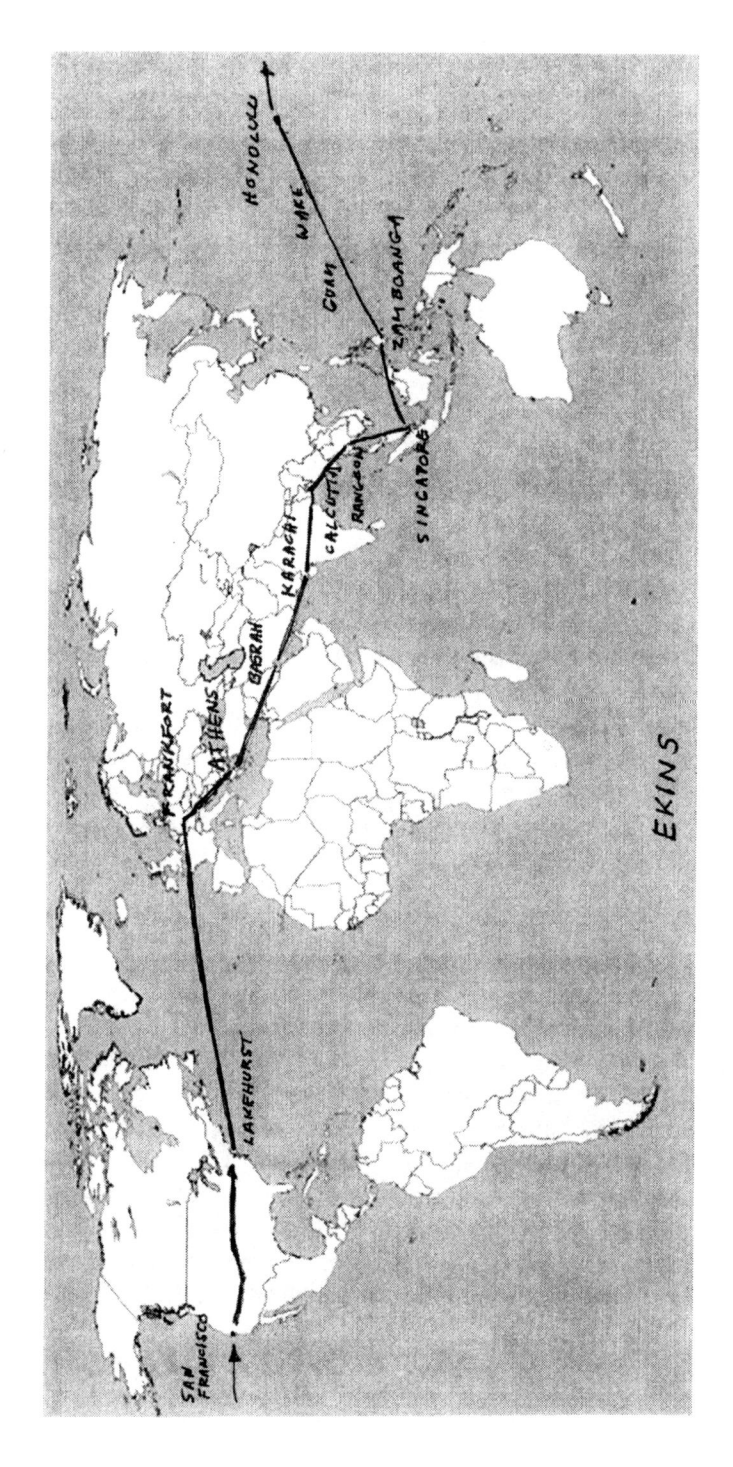

Circumnavigation

CHAPTER 12 – GALLAGHER

Two fast trips followed Ekins' run, but neither qualified as a full circumnavigation. Howard Hughes and four assistants completed a circuit in under four days in 1938. They traveled a short 15,000 mile route similar in distance to the trips of Wiley Post. In 1939, a woman lowered Elkins record one more time before the world was disrupted by war. She was known as a "first flighter." Mrs. Clara Adams had been on the first flight of the Graf Zeppelin in 1928, the Dornier D0-X flying boat in 1931, the Hindenburg in 1936, and the China Clipper flying boat, also in 1936.

Now, she would once again be a passenger on a first flight, Pan Am's Dixie Clipper flying boat, which left Port Washington, Long Island on its way to Horta, in the Azores on June 23, 1939. Mrs. Adams followed a route that took her to Lisbon, Marseille, Leipzig, Athens, Basra, Jodhpur, Rangoon, Bangkok, Hong Kong, Manila, Honolulu, San Francisco, and Newark. She rode scheduled airlines all the way, and her 25,000 mile trip cost $2,500 or $ .10 per mile. Forty years later, passage on a special Pan Am flight around the world cost $1,865, coach accommodation. Aviation reduced the cost of travel substantially as technology grew.

There is no indication that Mrs. Adams returned to Port Washington, Long Island, or beyond from Newark to complete the voyage.

Ekins' record continued to stand until after World War II for distance and a true, documented record.

Circumnavigation

During the war, transport and military planes grew larger and larger. By 1945, the Douglas C-54 Skymaster was one of the mainstays of the Army Air Force. Powered by four reciprocating engines driving big propellers, the C-54 was pulled through the air at nearly 300 miles per hour, while carrying 40 passengers. This popular aircraft was a natural for inaugurating the U. S. Army Air Transport Command service around the world on a weekly schedule. On a set timetable, the USATC would cover the 23,000 mile route starting each Friday and finishing the following Thursday night, at a cost of $2,341, plus $351.15 tax.

When the C-54 took off on the inaugural flight from Washington National Airport on September 28, 1945, it carried the name "Globester" and eight passengers. They were three reporters, four Air Transport Command Officers, and a War Department photographer. Another passenger was to disembark in Calcutta. In addition, nearly 3,000 pounds of mail and cargo were on the flight. Unlike other world trips, this one was not a single stunt, but the beginning of regularly scheduled air service linking the farthest reaches of the earth for military personnel.

The Globester, after taking off eastbound at 3:58 PM, Central War Time, made its first refueling stop at Kindley Field, Bermuda, continuing from there to the Azores at 9:30 PM. After stopping at Santa Maria, the southernmost island in the Azores, she made a short hop to Casablanca. A total of six C-54's would be used eventually to complete the circle, which covered a route stopping at Tripoli, Cairo, Abadam in Iran, Karachi, and Calcutta in India, Luleani near Kunming, China, Manila, Guam, Kwajalein, Johnston Island, Honolulu, and San Francisco.

The passengers flew over the Himalayas at 14,500 feet to have breakfast at Kunming, China, lunch in the battle scarred ruins of Manila, where the Globester picked up 33 homeward bound prisoners of war, and had a box lunch supper at 10,000 feet between Manila and Guam, landing at 1:00 AM, after a seven hour flight from Manila.

Upon leaving Guam, the C-54 developed magneto trouble in the number two engine, forcing the plane to return to Guam. Captain Marion H. Glick of Denver decided to return to Guam after covering only 315 miles of the 1,551 miles to Kwajalein. This constituted the only significant problem in the 23,000 mile trip.

The passengers gained one day when crossing the International Dateline. The final leg of the flight consisted of a 2,500 mile nonstop flight coast to coast from San Francisco to Washington, D. C. The seven men and one woman on board landed at 8:42 PM, on October 4, 1945. They had covered 23,279 miles in six days, five hours, and 44 minutes, including 33 hours and 21 minutes of ground time.

Ekins' time had been bested by 13 days and the earth had shrunk to nearly a third of its 1936 size. One of the things which stood in the way of faster time was the 33 hours of ground time, a problem shortly to be eliminated by the Air Force and its large, and powerful, long range bomber.

James Gallagher had plenty of flying experience during the war. He had flown the Hump, the famous cargo run over Burma. He had been among the first Americans to bomb Japan after Doolittle. He had worked on atom bomb tests. Captain Gallagher had flown B-29's in Germany and Alaska and trained extensively for a special secret mission in the highly modified B-29, designated the B-50.

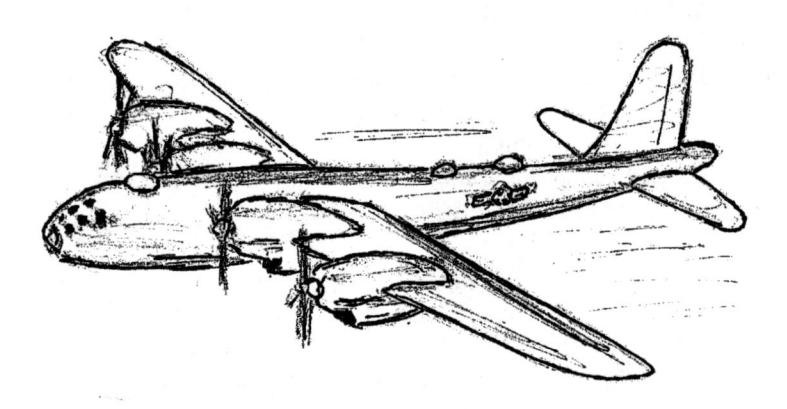

It was not until after thorough training in the air to air refueling techniques that the crews were briefed on their mission on February 21, 1949.

Gallagher and his crew were to be backup for an around the world flight without landing, to be performed by Lt. Forrest Jewell. Jewell's plane took off on February 25, and the dejected crew of the Lucky Lady II, Gallagher's B-50, could do nothing more than wish them well.

Jewell's plane had to abort the attempt in the Azores and Gallagher's Lucky Lady II was readied for the next try.

Seventy gallons of water and many cans of food were loaded onto the big bomber. Filled to overflowing, the gas tanks of the Lucky Lady brought her weight well above 142,000 pounds. Takeoff would be critical. Any mechanical failure in the heavily laden bomber could bring it down with its load of high octane. The big engines turned over and she taxied out for takeoff.

The B-50 was designed to carry bombs over enemy territory at high altitudes, like her predecessor the B-29. She was powered by four dynamic engines that would have amazed Ekins. As Gallagher's big propellers bit into the air, the Lucky Lady lumbered forward along the runway at Carswell Air Force Base at Fort Worth. Carefully, she was lifted into the air, and slowly Gallagher raised the wing flaps until the critical time had passed, and they climbed through overcast skies toward the stratosphere over Texas. Gallagher headed east toward the Azores. Although bigger than the aircraft of Elkin's time, the B-50 was not designed for comfort. A small wooden table was built into the forward pressurized cabin. Two men could sleep on the table on air mattresses. Food was stored below this table. A thirty foot tunnel connected the front cabin with the tail of the aircraft, and two men at a time slept in this tunnel over the bomb bays.

Food was prepared by opening the self-heating cans whenever members of the 14 man crew hungered. Although a cramped and grueling flight, Del Cano, or Drake's crews might have considered it the epitome of luxury.

The Lucky Lady left Fort Worth at 11:21 AM, Saturday, February 26, 1949. At 8:00 AM, Sunday morning, two tanker planes rendezvoused over the Azores. The refueling process required the Lucky Lady to trail out a haul line behind the fuselage. The tanker, an aviation gasoline carrying B-29, then fired a contact line which hooked the Lady's haul line. The haul line was then used to drag the heavy fuel line back to the Lady. Once attached, the planes flew nearly side by side while gasoline flowed from the tail of one to the tail of the other through a fuel line bowed behind them by the drag of the air.

The Lady almost suffered a casualty when a gunner nearly lost a finger in the mechanism. Quick action saved the finger, and with tanks topped off, the Lady roared eastward toward Gibraltar.

With the near loss of a finger being a recorded event in an around the world voyage, history had changed dramatically since the days when hundreds died before an expedition returned home.

The second refueling came off the east coast of Saudi Arabia, after an uneventful flight across barren North Africa. The Lucky Lady crossed India that Monday, flying over areas that were still wild and rugged. They passed the Andaman Islands where cannibals still lived. Over the Straits of Malacca, destination of many early circumnavigators, the B-50 passed through a squall with lightning streaking straight down to the earth.

The next refueling occurred over the Philippines, Tuesday morning. One of the planes, piloted by Captain Fuller, emptied its fuel load into the Lady, then headed home. Fuller's plane crashed somewhere in the area, a reminder of the uncertainties of travel in any age.

Hawaii was next. The Lucky Lady passed the dateline and it became Tuesday once again. As she approached the California coast, the B-50 flew through another thunderstorm with ice forming on the wings. Just twenty years earlier, such ice would have been extremely dangerous. Now, Gallagher had but to switch on the de-icing equipment. At 3:00 AM, Wednesday, March 2, they spotted the coast of California.

Over sunny Arizona, three tankers from Fort Worth met them to escort the exhausted flyers home. As the excited men approached Carswell Air Force Base, they realized that the auxiliary generator was out, causing a little longer than usual landing. Lt. General Curtis LeMay, standing in the crowd of dignitaries remarked, "He's been up there so long, he's forgotten how to land."

The Lucky Lady landed that March 2, where she had taken off just 94 hours, one minute, and 23,453 miles earlier. The earth had been chopped a little smaller once again.

The next major globe circling did not quite cover the full 22,859 mile length of the Tropic of Capricorn, but it was significant. It was accomplished not by an elaborate military machine, but by a young woman from Chicago. Her name was Pamela Martin, and she showed how far airline travel had developed since the time of H. R. Ekins.

The time was 1953, and the difference was a new engine that relied on Newton's third law of motion for its movement. The powerful jet engine was now capable of pushing aircraft through the air at over 600 miles per hour, nearly twice the speed that they could be pulled along by their propeller driven counterparts. The new jet aircraft could fly higher above the turbulent air and travel became even more comfortable. Although jets would not replace the commercial fleet for a few years, the first jetliners, the four engine British Comet, would speed Martin along her way, and her distance would be just barely short of the required record.

The 23 year old Martin was a native of New Mexico and was working in Chicago as an advertising copywriter. Her $1,751.70 fare was paid for by the Travel Agency of Journey-International Happiness Tours to prove that a woman could travel speedily around the world in comfort.

Starting December 4, 1953, the first leg of the journey was from Chicago to New York aboard a TWA Constellation where a ten minute helicopter ride took her from LaGuardia to New York's Idlewild International Airport. She boarded a British Overseas Airways Corporation (BOAC) Stratocruiser for London. In London, she had the longest stop of the trip, four and one half hours.

It was here that she boarded the BOAC Comet that would take her to Tokyo. At altitudes up to 39,000 feet and speeds up to 622 miles per hour, she traveled to Rome, Cairo, Bahrain, Karachi, Delhi, Calcutta, Rangoon, Bangkok, Manila, Okinawa, and Tokyo, with each stop being a brief one for refueling. From the air, she could see the deserts, mountains, and oceans that had been viewed by so many world travelers over the centuries.

At Tokyo, she transferred to a Canadian Pacific Airliner for Vancouver. By riding the jetstream and skipping the Anchorage, Alaska, refueling stop, the pilot hoped to reach Vancouver in 13 instead of 17 hours, but bad weather forced the Anchorage stop. In Vancouver, she took a United Airlines DC-3 to Seattle to pick up a United DC-6 for Denver, where a snowstorm forced a short delay.

The final leg by DC-6 was from Denver to Chicago where she arrived just 90 hours and 59 minutes after leaving. Now, civilian aviation had caught up with military aviation once again, and a young woman had again seen the world in the footsteps of Nellie Bly. Her reward for the grueling three and one half day speed record was a leisurely trip around the world, courtesy of the travel agency, and a footnote in history.

In the see-saw of technology, the next records would stay with the U. S. military, but only for a short time.

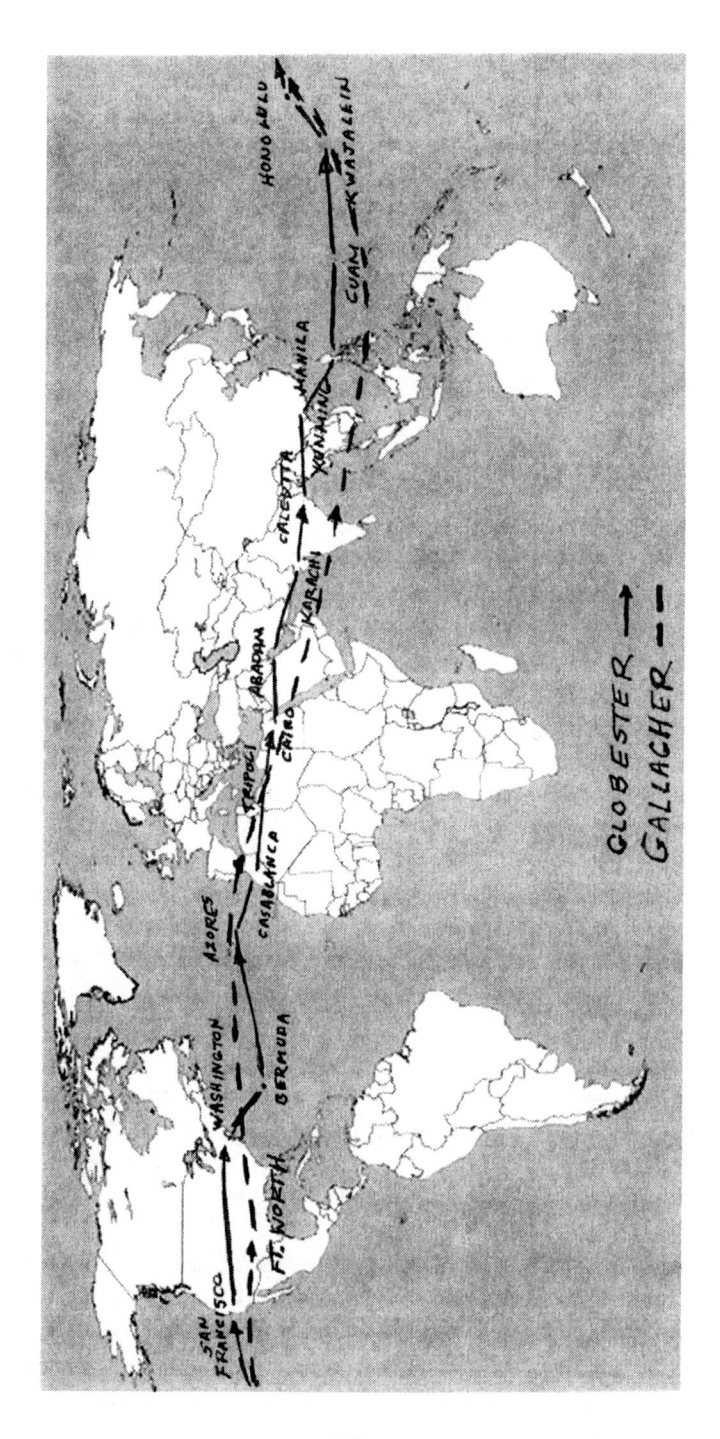

Circumnavigation

CHAPTER 13 - OLDS

The Lucky Lady II was both the beginning and the end of an era in military transportation. She was designed to carry the A-bomb and the H-bomb, the deterrent to war itself, and the technology that ended the previous war. In one form or another, war had sparked technology for some of the early circumnavigators as well as their later counterparts. The A-bomb era was just beginning, but the era of big propeller driven aircraft was ending. Big, slow planes with large wings for lift were soon to be replaced with sleek swept-wing jet aircraft with more power, more speed, and less wing.

In the nineteenth century, the fifties brought the rapid development of the clipper ship and travel time was cut in half. In the twentieth century, the fifties brought the rapid development of the jet airplane, and cut travel time in half again. Cruising speeds went from 300 to 600 miles per hour.

Following World War II, the world divided into the Communist World and the Free World, each armed with nuclear bombs and the means of delivery. For many years, the backbone of the United States long range bomber fleet was the B-52. Eight powerful jet engines drove these machines through the skies at over 40,000 feet.

On January 15, 1957, five of these aircraft took off from Castle Air Force Base at Merced, California, in a mock military mission. They were unarmed except for tail gun ammunition, and Russia was briefed in advance about the mission. Every detail was carefully planned. State Department clearance, worldwide communications, standby refueling planes, and precisely timed checkpoints were arranged. This kind of communication had been impossible for Magellan's starving crew, which traveled in a partially uncharted world.

Steak, canned chicken, cold milk, soup, fruit juice, cake, and candy sustained the crew that was to circle the globe in a time that would barely have allowed the early ships to have cleared port.

The mission was commanded by 50 year old Major General Archie J. Olds, Jr. The blue eyed Texan was a veteran of World War II, like Gallagher, and was now commander of the Strategic Air Command's Fifteenth Air Force. He stayed in radio contact with General Curtis LeMay's headquarters in Omaha, Nebraska, at all times. This demonstrated another wonder of contemporary communications, and brings to mind the days and ships lost by early voyagers separated by relatively short distances in bad weather with no good means of signaling each other. In the early days, if you were out of sight, you were lost, now you flew above the clouds talking to any receiver of choice the world over.

The airplanes thundered smoothly through the skies of North America, passing Newfoundland. One aircraft dropped out with mechanical failure over Labrador, and another landed in England as prearranged. Three continued onward over North Africa, Saudi Arabia, and Ceylon. They made a mock bomb run near the Malay Peninsula. They gave the Soviet Union a wide berth and, following the bomb run, crossed the vast Pacific, where Magellan's crew had lain weak and starving cut off from all they knew.

Four or five times they dropped to propeller altitudes and speeds to be refueled by KC-97 tanker planes. Unlike the Lucky Lady II, they drank directly from booms hanging behind the big four engine tankers, then climbed to more efficient cruising altitudes. At times, they traveled through air charged with static electricity, the wingtips glowing with St. Elmo's fire, the awesome phenomenon that stunned sailor and aviator alike over centuries of time.

On January 18, the three remaining aircraft passed Merced, California, and traveled to the March Air Force Base at Riverside, California. They had traveled 24,325 miles in 45 hours and 19 minutes, at an average speed of 550 miles per hour, with 27 crewmen. Without loss of life, mutiny, or minor difficulty, man had circled the globe in less than two days.

Civilian transportation would once again nearly catch up with military transportation in 1976, when Captain Walter H. Mullikin commanded a Pan American 747-SP from New York, to Delhi, to Tokyo, to New York in 46 hours and 50 seconds, a record for commercial aircraft. The flight covered 23,230.3 miles and suffered a short delay when a labor dispute slowed the Tokyo refueling stop.

The final record breaking circumnavigator would pass from land to sky, leaving stratosphere for ionosphere, and ionosphere for space in the ultimate attempt circle the globe. This time it would not be called circumnavigation, but orbiting. It would add new words to our vocabularies, and new dimension to our thinking.

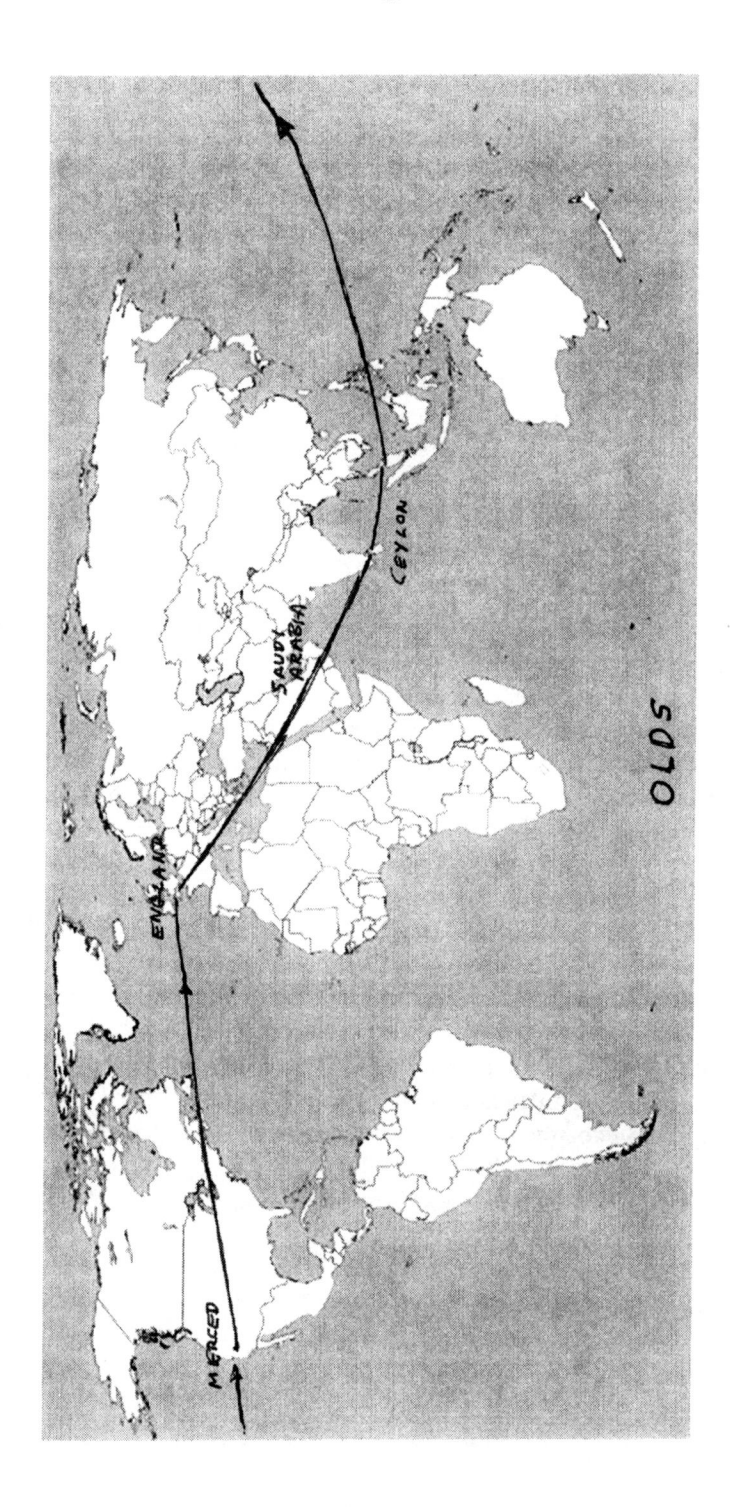

Circumnavigation

CHAPTER 14 - GAGARIN

"Visibility good! Can see everything! Certain spaces are covered with cumulus cloud formations."

A voice was coming to the earth from outer space. A man was circling the globe at nearly 18,000 miles per hour, and at an altitude reaching almost 190 Miles. It was April 12, 1961. The man was a Russian named Yuri Alekseyevich Gagarin.

For the first time, a circumnavigator could see the curvature of the earth beneath him, as he rode encapsulated in a tiny ship far above the atmosphere. Just as speed increased when man shook loose the bounds of the earth's surface to travel in the atmosphere, speed increased when he shook loose the bounds of the atmosphere to travel in space. Since the times of Francis Drake or John Byron, around the world treks had been a matter of intense nationalistic pride with military overtones, and this was no exception. The world was four years into the "space race" between the United States and Russia, a race which had begun on October 4, 1957, with the USSR launched Sputnik I.

Sputnik I beeped into the United States a new era of complete reevaluation of our status in the world. Still flushed with victory in a war that had been over twelve years earlier and blessed with the consumer prosperity of the fifties, the U. S. stood humiliated by the apparent technological superiority of its cold war rival. While Russia placed payloads of nearly 3,000 pounds in orbit, America's Vanguard rocket exploded in flames, unable to carry its 3.2 pound payload off the launching pad.

The very fabric of U. S. society was suddenly being redirected toward science and technology. Educational institutions almost instantaneously changed their approach. The race for space was on. Who would be the first to put a man in the outer reaches?

When this race ended with the flight of Vostok I, it was incidental that it was a single trip around the globe. While the first U. S. astronaut, Alan Shepard was scheduled for only a short, suborbital flight, Gagarin would have to make either one orbit or 17-18 orbits in order to return to the programmed landing area in the Soviet Union. This was because the earth would turn beneath the spaceship as it traveled in a fixed circle around the globe at an angle. The single orbit would take Gagarin across the equator twice at an angle of 65° and to the Arctic and Antarctic areas. It was close to a north-south orbit. When he returned, he was destined to pass the latitude of launching and land, because of the earth's movement beneath him, nearly a thousand miles west of the point of departure.

Gagarin had always wanted to fly. He had been but a child when the Germans had overrun his village, and he watched them pass by on their way to the front at Stalingrad. He went to technical school in Moscow, and in 1955, enrolled in flight school, where he learned to fly the Yak 18. He also learned parachute jumping. He continued in flying, joining the military and learning to fly MIG fighters. He married a nursing student named Valya, and they had two daughters by the time of his space flight. On his 26th birthday, he knew that his application for space flight training had been accepted, but he was unable to reveal his mission to Valya. Once he told her as a joke that he was on his way to outer space and to pack a clean shirt. She must have sensed the truth for she asked no more questions.

Gagarin and his backup pilot, Gherman Titov, who would become the second man to orbit the earth, spent the final hours before launch in a small brown cabin with walls of robin's egg blue. Gagarin spent these hours listening to soothing music, playing pool, and dining with the flight physician. Although Gagarin slept peacefully prior to launch, the designer of the spaceship, Sergei Korolev, could not sleep. More preparation had gone into the coming two hour flight of one man than had gone into the launching of Magellan's fleet over 400 years earlier.

The Russians have a saying, "The East is closer to the sun than the

West." Hence, the Russian name for east, Vostok, was selected for the world's first spaceship. In the desolate Kazakhstan Steppes, the 10,419 pound Vostok sat perched on top of the multi staged rocket which was to carry it aloft. On the morning of April 12, 1961, Senior Lieutenant Gagarin made a speech before being hoisted to the space capsule, 15 stories above the ground. He climbed into the capsule, and secured the hatch, thus encapsulating him in his tiny chamber. Then, he ran through some preflight procedures.

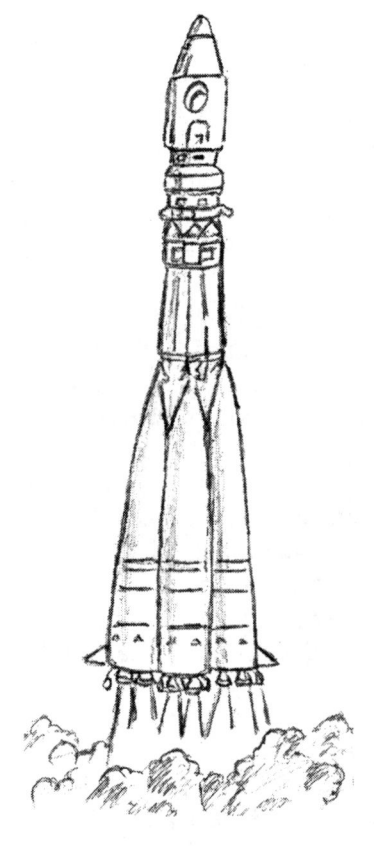

At 9:07 AM, the engines roared, and the big rocket began to lift off. Sealed inside the chamber, Gagarin felt the G forces of lift off, and heard the thundering of the engines. He still could not see out, since his portholes were covered with the nose cone which helped the rocket pierce its way through the atmosphere. The nose cone jettisoned itself in the upper atmosphere. When it popped off, the lieutenant couldn't conceal his excitement.

Circumnavigation

"How magnificent!" he exclaimed.

Below him were the cloud covered forests of his native country. The curvature of the earth could be seen, and he felt himself weightless as the centrifugal force of the orbit offset the gravitational pull.

The rockets had inserted the Vostok into an orbit at a 65° angle to the equator. Gagarin had left the "cosmodrome" at Baikonur at 47° 22' N latitude and 63° 25' E longitude. The ship passed over Siberia on the daylight side of the earth, then on toward the vast Pacific Ocean. The occupant of the ship could see the dividing line between day and night, a deepening blue line of lengthening shadows that turned into the blackness of night.

Vostok passed diagonally southward across the Pacific from west to east, covering the vast watery distance in a matter of minutes. At 9:52 AM Moscow time, just 45 minutes after liftoff, Gagarin reported that he was over Cape Horn, South America, where early circumnavigators had bitterly fought stormy weather on their way to the west.

At 10:15 AM, Major Gagarin, who had been promoted in flight, skipping the rank of captain, reported being over South Africa. At this point, the retro rockets were armed. The Vostok was turned to re-entry attitude. Riding the fine line between bouncing off of the atmosphere into space if the entry angle should be too shallow, and burning up in a meteoric flash if the entry angle were too steep, the Vostok would have to return at the perfect attitude to survive.

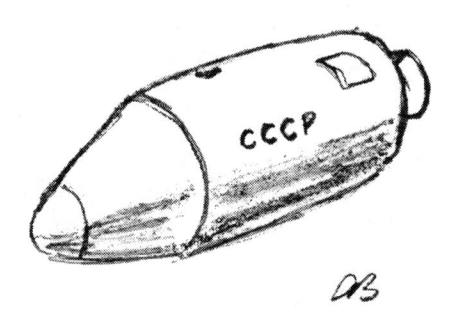

One hour and eight minutes into the flight, Gagarin was passing near Mt. Kilimanjaro in Africa. Minutes later, the retro rockets fired. Gagarin entered the most frightening part of the voyage. Through the portholes, he could see the reddish reflection as the Vostok's skin became hot and flames licked the sides. G forces pressed him into the contoured seat, and, dangerously, the spaceship began to tilt as it hit the first dense portions of the atmosphere. The ship righted itself, however, and slowed as intended.

The Volga passed beneath him. Gagarin had passed the area of lift off, but the earth had turned underneath that flight course in the last hour and three quarters, so that the spacecraft was destined to land far to the west of the original launch path. The round ship floated down beneath the red and white parachutes near the village of Saratov, not far from the programmed landing point. It was 10:55 AM.

The man from outer space surprised a woman and a girl who quickly approached him.

"Did you really come from outer space?" they asked.

Gagarin was an instant celebrity. His hero's welcome far exceeded the British welcome of Francis Drake's successful return. He had been the front man for a great accomplishment.

He also had left the ground and circled the earth before returning to the ground in an incredible one hour and 48 minutes.

In 442 years the world had shrunk from an uncharted journey of three years to less than two hours in circumference. Men had struggled with human and technological limitation to achieve this. They were Spaniards, Portuguese, Britain, Dutch, French, Americans, and Russians, and they brought the world together for better or for worse.

During this four and a half century period, the Spanish held the record one time, totaling 58 years. The British had nine records totaling 105 years, the Dutch two for 149 years, the French once for 50 years, the Americans eight for 80 years, and, as of 2012, the Russians for 51 years.

Who is likely to go faster, and why?

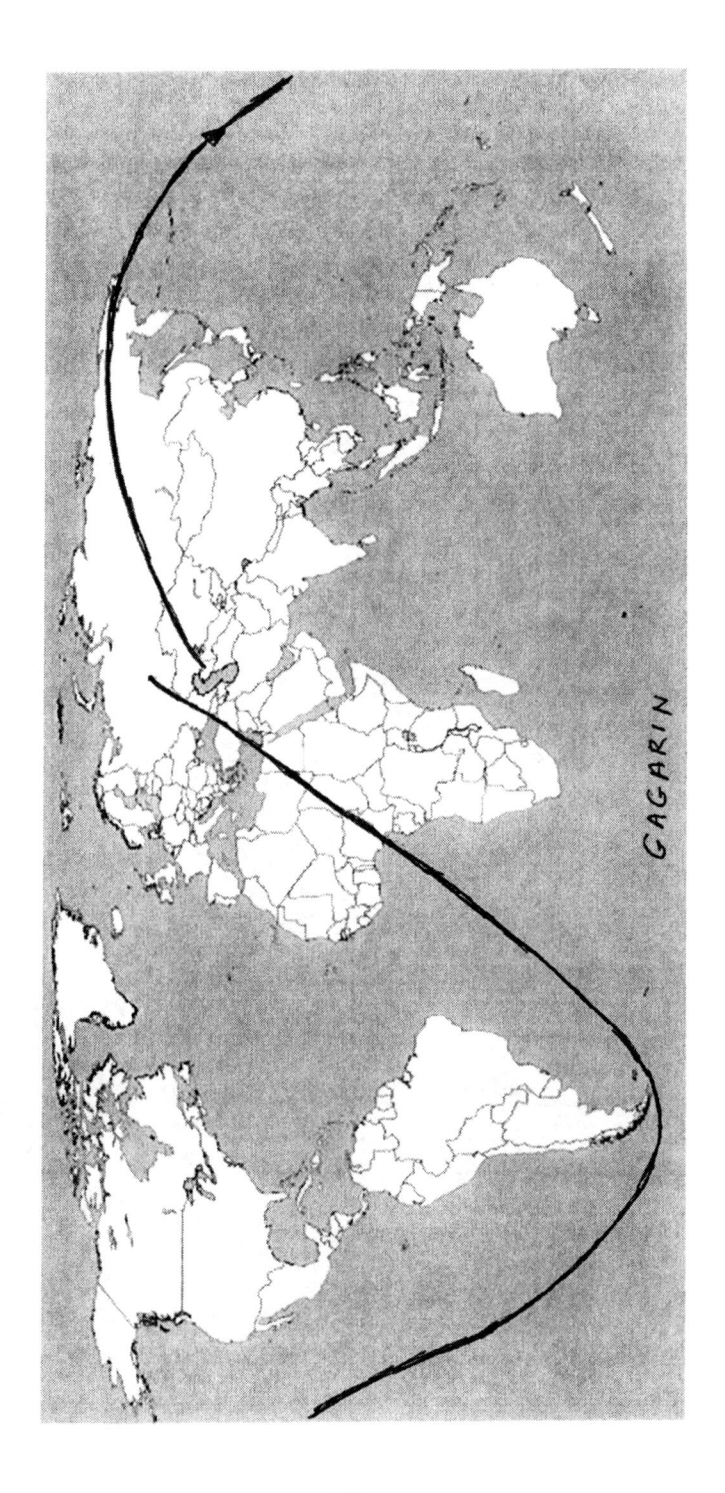

Circumnavigation

Circumnavigation Record Speeds

	Start Date	Voyage	Time
1	1519	Sebastian Del Cano – Victoria (Sp.)	2y, 11m, 17d
2	1577	Francis Drake – Golden Hind (Br.)	2y, 9m, 13d
3	1586	Thomas Cavendish – Desire (Br.)	2y, 2m, 10d
4	1615	Willem Schouten – Various (Du.)	2y, 0m, 17d
5	1721	Jacob Roggewein – various (Du.)	1y, 10m, 20d
6	1764	John Byron – Dauphin (Br.)	1y, 10m, 18d
7	1790	Etienne Marchand – Solide (Fr.)	1Y, 8m, 0d
8	1840	George Vasimer – Ann Mckim (Am.)	11m, 21d
9	1840	George Vasimer – Ann Mckim (Am.)	11m, 21d
10	1848	Bob Waterman – Sea Witch (Am.)	11m
11	1849	Capt. Sproat – Phoenician (Br.)	6m, 20d
12	1851	Capt. Sproat – Phoenician (Br.)	6m, 15d
13	1852	Bully Forbes – Marco Polo (Br.)	5m, 22d
14	1854	Samuel Heid – Red Jacket (Br.)	5m, 10d
15	1854	Bully Forbes – Lightning (Br.)	5m, 8d
16	1854	Capt. McDonnel – James Baines (Br.)	5m, 11d (1)
17	1889	Elizabeth Cochrane – various (Am.)	2m, 12d
18	1936	H. R. Ekins – various (Am.)	18d, 11h
19	1945	USATC Globemaster (Am.)	6d, 5h
20	1949	James Gallagher – Lucky Lady II (Am.)	3d, 22h
21	1957	Archie Olds – USAF B-52 (Am.)	1d, 21h
22	1961	Yuri A. Gagarin – Vostok I (USSR)	1h 48m

(1) See text

Index

"Bully" Forbes....85, 86, 87, 117

Aberdeen White Star Lines....84

Abyssinia, the 102

Alan Shepard 133

Andre Jaeger-Schmidt..........106

Ann McKim, the 82, 83

Architect, the...........................83

Augusta Victoria, the 93

B-29 ....122, 123

B-50 ....122, 123, 124

B-52 ....128, 138

Baab...42

Baikonur 135

Balboa, Vasco 11

Ball Line 81

Batavia, the 65, 71, 115

Black Ball Line 85, 86, 87

BOAC Comet.......................125

Burnley-Campbell................106

Byron, John....67, 68, 69, 70, 71, 75, 132, 138

C-54 121

Cacafuego 40

Campania, the 105

Cape Horn.......60, 70, 75, 77, 81, 82, 83, 84, 87, 135

Cape of Good Hope...10, 28, 44, 52, 55, 56, 60, 68, 77, 81, 84, 86, 90

Cape Verde Islands....16, 29, 35, 56, 68, 75

Capitana, the 39, 40

Captain Cook 75, 77

Captain Rondong 114

Captain Sproat.......................84

Captain Van Bremer 115

Cartagena, Juan...15, 16, 17, 18, 19, 20

Carvalho, Joao Lopez.............27

Castle Air Force Base 128

Cavendish, Thomas...47, 48, 49, 50, 51, 52, 53, 55, 69, 138

Charles B.D. Collyer............108

China Clipper.......................120

Christopher, the...10, 33, 34, 35, 36, 37, 93

Clara Adams 120

Cochrane, Elizabeth 91, 138

Columbus, Christopher 10, 12

Concepcion, the...15, 19, 21, 22, 23, 49

Consolidated National Company............................56

Content, the 48, 50, 51

cosmodrome.........................135

Curtis LeMay 124, 129

D0-X 120

Daniel DeFoe 63

Darwin, Charles 80

Dauphin, the.............68, 71, 138

DC-2.....112, 113, 115, 116, 117

DC-3 ....113, 125

DC-6 125

de Coca, Antonio 17, 18

de Silva, Nino 35, 36

Del Cano, Sebastion....14, 19, 20, 28, 29, 35, 44, 45, 123, 138

Desire, the......48, 50, 51, 52, 69, 138

Deutschland, the...................105

Don Manuel Quezon............115

Donald McKay.................86, 87

Doughty, Thomas......33, 34, 35, 37, 38, 43

Drake, Francis.....31, 32, 33, 34, 35, 36, 37, 38, 39, 40, 41, 42, 43, 44, 45, 47, 48, 50, 51, 52, 53, 55, 58, 75, 117, 123, 132, 136, 138

Dum Dum 114

Dutch East India Company... ....56, 65

Dutch West India Company...62

Easter Island.....................64, 70

Edward S. Evans..................107

Eendracht, the......56, 57, 58, 59, 60, 71

Ekins, H. R. 109, 110, 111, 112, 113, 114, 115, 116, 117, 118, 120, 122, 123, 125, 138, 145

El Paso.............18, 19, 21, 22, 25

Elizabeth, the..33, 35, 36, 37, 38, 49, 51, 91, 138

Enrique.................14, 25, 26, 27

Eratosthenes............................9

Espinoza, Gonzolo 28

Falkland Islands 68, 69

Falliero, Ruy 12, 13

Federation Aeronautique International.........................7

Fitzmorris, Charles...............104

Forbes, "Bully"...85, 86, 87, 138

Francis Fletcher................34, 43

Francis Petty 47, 49

Franklin D. Roosevelt 115

Fulton, Robert 81

Gagarin, Yuri......132, 133, 134, 135, 136, 138

Gallagher, James........122, 123, 124, 129, 138

General Werder....................102

George Thomas......................86

Globester..............................121

Golden Hind, the..12, 33, 38, 39, 40, 41, 42, 43, 44, 45, 138

Graf Zeppelin.......................108

Harold Getty 108

Hawaii Clipper.....................116

Henry Frederick...................105

Henry Mears 106, 107

Henry Rosenthal 63

Herodotus................................9

High Gallant, the....................48

Hindenburg, the...110, 111, 112, 120

Hoorn, the...55, 56, 57, 58, 59, 60, 70

Hougua, the............................83

Howard Hughes 120

Hugh Gallant, the 48, 49, 50

Hyades, the 105

Island of the Flies 58

Isthmus of Panama 11, 91

James Baines, the 85, 86, 87, 88, 91, 138

James Bauman 63

John Cockerill 92

John Fitch 81

John Fry 34

John Hawkins 32

John Thomas 33

Juan Fernandez Island 63, 70

Judith, the 32

Jules Verne 90, 91, 93, 94, 98, 99

KC-97 129

Kilgallen, Dorothy 111, 114

King Charles 12, 13, 15, 27

KLM 112, 113, 114

Korolev, Sergei 133

Lakehurst 108, 110, 111, 114, 117

LeMaire, Isaac 56, 57, 58, 59, 60, 62, 63, 71

Linton Wells 107

Liverpool 84, 85, 86, 87, 88, 104, 106

Lord Byron 67

Lucky Lady, the 122, 123, 124, 128, 129, 138

Lucky Lady II, the 122, 123, 128, 129, 138

Magellan, Ferdinand ... 12, 13, 14, 15, 16, 17, 18, 19, 20, 21, 22, 23, 24, 25, 26, 27, 28, 29, 31, 32, 33, 35, 37, 38, 39, 41, 42, 45, 47, 48, 49, 52, 56, 58, 59, 60, 62, 67, 68, 69, 75, 92, 95, 98, 117, 129, 133

March Air Force Base 130

Marchand, Etienne 72, 74, 75, 76, 77, 78, 80, 82, 138, 146

Marco Polo, the ... 85, 86, 87, 88, 138

Marigold, the ... 33, 35, 36, 37, 38

Marion H. Glick 121

Mary, the 35, 36, 37, 49

Mauretania, the 107

Maury, Matthew...............80, 85

Mayor LaGuardia.................115

Melbourne, the.....84, 85, 86, 87, 88

Memnon, the 83

Mendoza, Luiz.....15, 17, 19, 20, 21, 76

Mesquita, Alvaro..18, 19, 21, 22, 23

Mogador..........................34, 39

Moluccas..........................10, 65

Morning Star, the 59

Nellie Bly......91, 92, 93, 94, 95, 96, 97, 98, 100, 102, 103, 109, 113, 117, 126

New York Herald...................83

Nuestra Senora, the 40

Oceanic, the 97, 98

Olds, Archie 129, 138

Oriental, the....83, 94, 95, 96, 97

Pamela Martin......................124

Pan American.......112, 116, 130

Pan American 747-SP 130

Patagonians...22, 37, 69, 70, 118

Pelican, the....33, 34, 35, 36, 37, 38

Phineas Fogg 90, 94, 99, 100

Phoenician, the...........9, 84, 138

Pigafetta, Antonio......14, 15, 17, 19, 20, 24, 26, 29, 31, 49

Plymouth ..33, 34, 37, 48, 56, 68

Pope Leo X 11

Port Desire............36, 49, 57, 69

Port Famine.....................69, 70

Pratt & Whitney 116

Prince Henry 10, 11

Pythagoras...............................9

Quesada, Gaspar...15, 17, 19, 20, 21

Rainbow, the 81, 83

Red Jacket, the....86, 87, 88, 138

Renege, Captain.....................32

Rio de Janeiro 17, 62, 63

Rio de la Plata 13, 18

Robinson Crusoe....................63

Roggewein, Jacob......62, 63, 64, 65, 67, 71, 80, 138

Samuel Reid...........................86

Samuel Russel.......................83

San Antonio, the...15, 16, 17, 18, 19, 20, 22, 23

San Julian 21, 28, 37, 43

San Salvadore 32

Santa Anna, the 51

Santa Maria, the 35, 121

Santiago, the...15, 18, 20, 21, 22, 28

Savannah, the 81

Sayre, J. Willis 104, 105, 106

Schouten, Willem.....56, 57, 58, 59, 60, 62, 64, 67, 71, 80, 138

Scripps-Howard Newspapers 110

Sea Witch, the 83, 138

Selkirk, Alexander 63, 70

Serrano, Juan......15, 17, 18, 21, 22, 23, 27

Sierra Leone 45, 48, 57

Solide, the.....75, 76, 77, 82, 138

Spice Islands...10, 13, 23, 25, 28, 31, 42, 52, 59, 63

Spilbergen, Joris...............59, 60

Sputnik.................................132

Strait of Magellan 23, 56

Suez Canal........77, 88, 100, 104, 113

Sutter's Mill 83

Swan, the...............33, 35, 36, 37

Tamar, the 68, 71

Tchinkitane 76

Tenerife.....................15, 16, 17

Terra Australis......18, 56, 59, 84

Tienhoven, the............63, 64, 65

Tierra Del Fuego....................58

Titov, Gherman....................133

Train, George Francis...102, 103, 104

Trans-Siberian Railroad............105, 106

Trinidad, the...12, 15, 16, 17, 18, 19, 20, 22, 23, 27, 28

Tropic of Capricorn...7, 104, 124

Vasimer, George 82

Victoria, the...15, 17, 20, 21, 22, 23, 28, 29, 33, 45, 93, 95, 96, 107, 108, 138

Vladivostok..................106, 107

Vostok I 133, 138

Walter H. Mullikin...............130

Waterman, "Bob"...........83, 138

Wiley Post....................108, 120

William Pickman 48

Woodes Rogers......................63

Wynter, John 33, 34, 37, 38

Yak 18..................................133

Zeppelin 108, 120

Bibliography

Cutler, Carl C.; *Greyhounds of the Sea*; Naval Institute Press; 1960 ISBN 087021232X

Dixon, George; *Voyage Round the World*; 1968 reprint of 1789 Edition ISBN 0-306-77133-0

Ekins, H.R.; *Around the World in 18 Days and How to Do It*; Longmans; 1936

Fairburn, William; *Merchant Sail*; Fairburn Marine Educational Foundation; 1945

Fleurieu, Charles P. C.; *A Voyage Round the World 1790-1792 Performed by Etienne Marchand*; London; 1801; ISBN 0-306-77121-7

Howe, Octavius & Matthews, Frederick; *American Clipper Ships 1833-1858*; Salem, Mass.; Marine Research Society; 1936

LaGrange, Helen & Jacques; *Clipper Ships of America and Great Britain, 1833-1869*; New York; G.P. Putnam's Sons; 1936

Lubbock, Basil; *The Colonial Clippers*; Glasgow, Brown & Son, 1924

McKay, Richard C.; *Some Famous Sailing Ships and Their Builder Donald McKay*; New York; G.P. Putnam's Sons; 1928

Silverberg, Robert; *The Longest Voyage*, Indianapolis-New York; Bobbs Merrill; 1972

Wilson, Derek, *The World Encompassed*, London; Allison & Busby LTD, 1998

ABOUT THE AUTHOR

Alan Boone lives in Indianapolis, Indiana and is a graduate of Wabash College, where he majored in Economics. He worked in the airport industry in Indiana and North Carolina. He is an Accredited Airport Executive (A.A.E.) and retired from forty years in the airport industry in 2011. During most of that time, he served as CFO for the Indianapolis Airport Authority. His fascination with transportation, people, politics, and technology led to this work. He and his wife Sandy have three children, and eight grandchildren.

CPSIA information can be obtained at www.ICGtesting.com
Printed in the USA
BVOW05s1958281114

377156BV00015B/698/P